Moritz Peter

Towards magnetic resonance in scanning tunneling microscopy using heterodyne detection

Experimental Condensed Matter Physics
Band 18

Herausgeber
Physikalisches Institut

Prof. Dr. Hilbert von Löhneysen
Prof. Dr. Alexey Ustinov
Prof. Dr. Georg Weiß
Prof. Dr. Wulf Wulfhekel

Eine Übersicht aller bisher in dieser Schriftenreihe erschienenen Bände finden Sie am Ende des Buchs.

Towards magnetic resonance in scanning tunneling microscopy using heterodyne detection

by
Moritz Peter

KIT Scientific Publishing

Dissertation, Karlsruher Institut für Technologie (KIT)
Fakultät für Physik
Tag der mündlichen Prüfung: 15. Mai 2015
Referenten: Prof. Dr. W. Wulfhekel, Prof. Dr. B. Pilawa

Impressum

KIT Scientific Publishing

Karlsruher Institut für Technologie (KIT)
KIT Scientific Publishing
Straße am Forum 2
D-76131 Karlsruhe

KIT Scientific Publishing is a registered trademark of Karlsruhe Institute of Technology. Reprint using the book cover is not allowed.

www.ksp.kit.edu

Print on Demand 2015

ISSN 2191-9925
ISBN 978-3-7315-0410-8
DOI 10.5445/KSP/1000048009

Contents

1 Introduction

The realization of the scanning tunneling microscope (STM) in 1982 by G. Binnig and R. Rohrer revolutionized the field of surface science [1]. It allows to explore metallic surfaces on the atomic scale, enabling humans to observe single atoms and molecules [2–4]. This ability made the STM a widely used tool in fundamental and applied research [5]. Especially the modern world of telecommunications depends on the continuous miniaturization of integrated circuits and storage units. During the development of such new applications, basic physical studies are necessary as the sizes approach limits where classical descriptions often fail and quantum mechanics determine the behavior. Here regularly, the STM is the tool of choice for studies of functional electronic and magnetic properties. In particular spin-polarized STM should be emphasized, which massively evolved in the last decade. It allows to detect the local spin polarization on the atomic level [6,7], which is the foundation of all modern spintronics.

But the STM displays one profound weakness and that is its limited time resolution. This originates from technical limitations of the current amplifier which detects the tunneling current of just some hundred pico to a few nanoampere. Even the best available current amplifiers are limited to several kilohertz. Thus, the measurement of dynamics in an STM is very limited. This is especially true for the dynamics of magnetism, respectively spin-related studies. There the timescales rarely exceed the nanosecond range. Thus, only static magnetic phenomena can be observed.

The contrast is given by magnetic resonance techniques. They probe the magnetism of macroscopic samples with photons of the gigahertz range [8]. These techniques are therefore ideally suited for studying the dynamics of spin ensembles. Their most prominent application is, of course, nuclear magnetic resonance tomography, where the different spin dynamics of organic materials are used to image the human body. But when it comes to the detection of very small samples and finally a single spin, magnetic resonance is limited by its signal detection technique.

The present PhD work introduces a new approach of combining STM and magnetic resonance. This should yield the best of both worlds, the excellent lateral resolution of STM and the time resolution of magnetic resonance. Hence, the aim is to enable measurements of magnetism dynamics on the nanometer scale. This starts with ferromagnetic resonance of nanoscale particles, where the dynamics are essentially determined by magnon-modes, and ends at single spins, where the states are split by external fields and anisotropies.

This prospect is especially attractive since the dynamics of nanometer-sized magnetic objects irresistibly develops towards quantum computation in current research [9–12]. Since the basic building block of a future quantum computer, a qubit, must be a quantum-mechanical two-level system, a single spin-$^1/_2$ is one possible means. But the realization of operations needed in a quantum computer requires the control and readout of this dynamically evolving system. It should be noted that the probably most successful implementations of basic quantum computer algorithms were realized by nuclear magnetic resonance on an ensemble of paramagnetic molecules [13, 14]. But therefore, pulsed radio-frequency photons are needed to manipulate the spins. This is exactly the requirement that conventional STM cannot fulfill. The drawback of the magnetic resonance approach is that it cannot be scaled easily to allow computation of many qubits.

Then again, the STM is already able to detect the time averaged contributions of a single atom or molecule spin. And since every spin can be identified individually, the problem of scalability is less severe. It is consequential that a measurement of the time evolution, and later maybe the manipulation of such quantum mechanical objects in an STM, could significantly stimulate this field.

With the general focus on measuring magnetic dynamics on the nanosecond timescale in single nano-objects, this work presents the development of a new experimental approach. Therefore the theoretical background of magnetic resonance and STM are explained initially in chapter 2. This is necessary to introduce the reader to the fundamental concepts of these two individual techniques. Afterwards, in chapter 3, the new technique of heterodyne detection of magnetic resonance signals in an STM (which is the basis of this work) is developed theoretically. Additionally, previous approaches are reviewed, which ranks and validates the present work. To realize the introduced technique, an STM in ultra-high vacuum at low temperatures was used. But since the setup exhibited a layout incompatible with magnetic resonance, a modification of the latter was carried out during the period of this PhD. Therefore, the whole chapter 4 is dedicated to the experimental setup, its modification for radio frequencies, and the characterization and revision of the latter. Following this, the first experiments are shown. Since the measurements are based on a totally new concept, its validity has to be proven. Thus, rather simple, well-understood proof of concept samples are needed. The first choice were iron islands on tungsten, which are presented in chapter 5. Although this sample does not tackle the target of single spin detection in an STM, it allows initial studies utilizing the new experimental technique. Subsequently, in chapter 6 and 7 the studies on two advanced systems consisting of magnetic skyrmions and paramagnetic molecules are presented. In the vicinity of this experimental development these preliminary studies have to be considered as proof of concept experiments, too.

2 Theoretical background

In this chapter the theoretical background of this work is explained. As the present work is based on the combination of two well established techniques, namely magnetic resonance, i.e. electron spin resonance (ESR) / ferromagnetic resonance (FMR), and scanning tunneling microscopy (STM), both are introduced separately. This will point out their strengths and also their disadvantages. A good understanding of both techniques and their different facets is important to realize their integration.

2.1 Conventional magnetic resonance

Conventional magnetic resonance techniques are based on the absorption of electromagnetic radiation by matter. As the term *magnetic* resonance already implies, the principle is founded on the interaction of radiation with magnetic moments of the sample. To access energy levels of different magnetic states, radio frequency- respectively microwave-sources and spectrometers are needed. Historically they were not tangible until Second World War. At that time, with the inducement of radar applications, these utilities became available. In 1944, the resonance of paramagnetic electron spins was observed for the first time by Zavoisky [15]. Only two years later, in 1946, Griffiths and Zavoisky independently discovered the resonance phenomena in ferromagnetic nickel [16]. In the same year the field of nuclear magnetic resonance arose. In contrast to the first two discoveries, which target the magnetic moment of electrons, nuclear magnetic resonance (NMR) addresses the magnetic moment of the nucleus. For the purpose of this work only the first two are of interest, because in the combination with a scanning tunneling microscope only electrons are accessible.

All these magnetic resonance methods are able to measure various properties of the sample material and its internal magnetism. Starting from simply unveiling magnetic energy level [8] through to imaging techniques [17], a lot of basic research is founded on magnetic resonance. Especially in chemistry, for structural analysis, NMR is a well established tool. But there are also highly specialized works, where simple quantum computation was realized (NMR) [13, 14] or proposed (ESR) [18]. This is possible since magnetic resonance allows investigation and manipulation of the dynamics of magnetic moments, respectively quantum objects. This access to the dynamics of magnetic matter is the most attractive aspect regarding the use in this work.

The two phenomena considered, electron spin resonance (ESR) and ferromagnetic resonance (FMR), are very similar up to the different materials probed and their fundamental magnetic properties. In ESR electron spins or orbital momenta of isolated atoms or weakly coupled paramagnetic systems are the target. FMR in contrast, probes strongly interacting exchange coupled electron systems. This exchange interaction does not directly influence the resonance of a totally magnetized sample but it creates an internal magnetic field, which can change the resonance condition significantly in comparison to ESR.

For simplicity and completeness the basic quantum mechanical effects which underlie ESR are introduced first. The second part will advance towards FMR, where all quantum mechanical effects are summarized to an effective field and the magnetic moments are combined to a magnetization.

2.1.1 Electron spin resonance

For a simple understanding of ESR the spin magnetic moment of a free electron has to be considered:

$$\boldsymbol{\mu}_{\mathrm{S}} = -g_{\mathrm{S}}\mu_{\mathrm{B}}\mathbf{S} = -g_{\mathrm{S}}\frac{e\hbar}{2m_{\mathrm{e}}}\mathbf{S}. \tag{2.1}$$

Where g_s is the dimensionless g-factor and μ_B the Bohr magneton. This magnetic moment occurs in the simplest form of the spin Hamiltonian, which consists only of the electron spin Zeeman interaction:

$$\mathcal{H}_{\mathrm{EZI}} = -\boldsymbol{\mu}_{\mathrm{S}} \cdot \mathbf{B} = g_{\mathrm{S}}\mu_{\mathrm{B}}\mathbf{S} \cdot \mathbf{B}. \tag{2.2}$$

It describes a free electron spin in an magnetic field $\mathbf{B}$. For a field in z-direction ($\mathbf{B}_0 = \hat{\mathbf{z}}B_{\mathrm{z}}$) one finds $2S + 1$ equally spaced eigenstates with energies:

$$E = g_{\mathrm{S}}\mu_{\mathrm{B}}B_{\mathrm{z}}m_{\mathrm{S}}, \quad m_{\mathrm{S}} = S, S-1, \ldots, -S. \tag{2.3}$$

It is obvious that the energy splitting between the states depends linearly on the external magnetic field. In a simple spin-$1/2$ system this yields two pure states, the spin-up state $|\uparrow\rangle$ and the spin-down state $|\downarrow\rangle$. The total wavefunction of such a system is then a superposition of all pure states weighted with complex coefficients α and β:

$$|\psi\rangle = \alpha\,|\uparrow\rangle + \beta\,|\downarrow\rangle \quad \longrightarrow \quad |\psi\,(t)\rangle = \cos\left(\Theta/2\right)|\uparrow\rangle + e^{i\varphi}\sin\left(\Theta/2\right)|\downarrow\rangle\,. \tag{2.4}$$

The transformation in equation 2.4 can be made because the full state needs to be normalized ($\langle\psi|\psi\rangle = 1$). The two remaining variables θ and φ can be used to describe a vector in a sphere. This is called the Bloch sphere [19] (fig. 2.1). The cosine (sine) of $\theta/2$ indicates the probability to find the system in the $|\uparrow\rangle$ ($|\downarrow\rangle$) state. These probabilities are constant in time for this simple situation. The variable φ, which is a relative phase, in contrast is not constant. It is found by solving the time dependent Schrödinger equation:

$$\varphi\,(t) = \varphi_0 - \frac{(E_\uparrow - E_\downarrow)}{\hbar}t = \varphi_0 - \frac{g_s\mu_B B_z}{\hbar}t \tag{2.5}$$

This linear relation can be identified as a precession of the state vector around the z-axis, the so-called Larmor precession (fig. 2.1(*a*)). The frequency of this precession is given by $\omega_L = g_S\mu_B B_z/\hbar$.

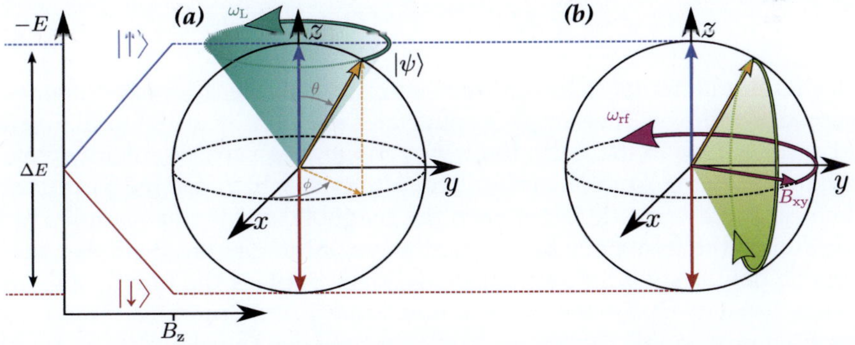

Figure 2.1: *At zero field the two pure states of a spin-½ system are degenerate. For none zero magnetic fields the states split up into $|\uparrow\rangle$ and $|\downarrow\rangle$ state. The dynamic behavior of the system is depicted in Bloch's sphere representation. Subfigure* ***(a)*** *shows the Larmor precession of the state $|\psi\rangle$ around the z-axis of the coordinate system. In* ***(b)*** *a spin-flip excitation induced by a small time dependent field (B_{xy}) in the xy-plane is shown.*

For spectroscopic investigation, like ESR, an interaction that triggers transitions between the levels is needed. This can, for example, be a small oscillating magnetic field in xy-direction:

$$\mathbf{B}_1(t) = B_{\mathrm{xy}} \left(\hat{\mathbf{x}} \cos(\omega_{\mathrm{rf}} t) + \hat{\mathbf{y}} \sin(\omega_{\mathrm{rf}} t) \right). \tag{2.6}$$

This perturbation of the Hamiltonian renders the absolute values of the coefficients α and β time-dependent, and therefore allows transitions between the states. Figure 2.1*(b)* shows the resulting spin flip excitation of the state vector in the Bloch sphere induced by the oscillating field. In this simple picture the perturbation can be understood as a circulating field. In resonance the Larmor frequency and the frequency of the circulating field are the same, thus in a rotating coordinate system the field in xy-direction causes a continuous swap between the two states. The energy of the interaction $\hbar\omega_{\mathrm{rf}}$ must equal the energy difference between the initial and the final state to fulfill energy conservation. The easy resonance condition in ESR is then given by:

$$\hbar\omega_{\mathrm{rf}} = \Delta E = g_{\mathrm{S}} \mu_{\mathrm{B}} B_{\mathrm{z}}. \tag{2.7}$$

The shown transitions are called magnetic dipole transitions and imply inherent selection rules of $\Delta m_{\mathrm{S}} \pm 1$ [20]. In a more general system, with many states, the transition rates and thereby the selection rules for an arbitrary interaction V_{int} are given by Fermi's golden rule [21], with $|i\rangle$ and $|f\rangle$ being the initial and the final states:

$$P_{i \to f} = \frac{2\pi}{\hbar} \left| \langle f \left| V_{\mathrm{int}} \right| i \rangle \right|^2 \delta \left(E_i - E_f - \hbar\omega_{\mathrm{rf}} \right). \tag{2.8}$$

With this fundamental quantum mechanical principle, ESR experiments are designed as follows: The sample's spin states are split by an external magnetic field and photons in the radio frequency range, respectively microwaves, are irradiated into the sample cavity. The photons, which are the perturbation as before, cause transitions between the magnetic states of a sample. In this environment the frequency of the microwave or the magnitude of the external magnetic field is swept. Simultaneously, absorption of the microwave inside the sample cavity due to the allowed transitions is measured. At resonance the photon energy equals the energy splitting between the samples' states, thus the absorption is maximal. This increase of the absorption can be detected. Consequently each allowed transition in the sample will yield a resonance peak. This is called continuous wave (CW) ESR.

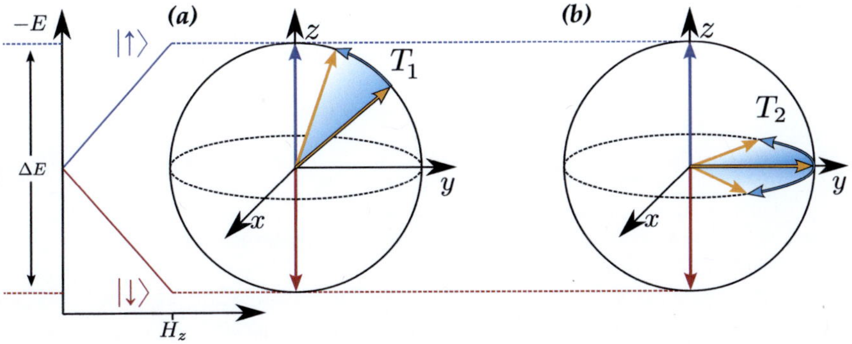

Figure 2.2: *The same situation as in figure 2.1 is present. Part **(a)** illustrates the longitudinal relaxation towards the energetically favored ground state. T_1 is the associated longitudinal relaxation time. In **(b)** the relaxation due to dephasing is sketched. The characteristic time T_2 is called transversal relaxation time.*

The given model is of course too simple for real problems targeting electrons in molecules or solids. In this environments the time evolution, i.e. the dynamic, is modified. In real systems relaxation mechanisms are present, which damp the resonance and causes a broadening of the resonance lines. The easiest analysis of the latter follows Heisenberg's uncertainty relation $\Delta E \geqslant \hbar\tau^{-1}$ [22]. In an CW-ESR experiment the lower limit of the lifetime τ of a state can therefore directly be obtained from the width (ΔH or $\Delta\omega_{\mathrm{rf}}$) of a resonance line:

$$\Delta B_{\mathrm{z}} \geqslant \frac{\hbar}{g_{\mathrm{s}}\mu_{\mathrm{B}}}\frac{1}{\tau}; \quad \Delta\omega_{\mathrm{rf}} \geqslant \frac{1}{\tau}. \tag{2.9}$$

For a more precise description of the relaxation(-mechanisms) and dynamics of a system, it is common to introduce two different relaxation times. Energy relaxation mediated by e.g. lattice vibrations, is usually introduced as T_1. It describes the tendency of the system to relax back into its energetic ground state after an excitation (fig. 2.2*(a)*). In the case of a spin bath this means relaxation into the thermal equilibrium given by the corresponding thermal distribution. The second relaxation time T_2 can be motivated through dephasing (fig. 2.2*(b)*). This can be understood when fluctuations in the phase velocity are considered. These can be caused by, e.g. interaction with neighboring spins or inhomogeneous fields. Because of the different phase velocities the expectation value of a spin in xy-direction will vanish over time. In most cases the dephasing time T_2 is much shorter than the energy relaxation T_1.

A general behavior of these relaxation mechanisms is covered by the classical Bloch equation, which is widely used in conventional ESR to explain time-dependent phenomena [19]:

$$\frac{d\mathbf{M}}{dt} = \underbrace{\frac{g\mu_\mathrm{B}}{\hbar}(\mathbf{M}\times\mathbf{B})}_{\text{precession term}} - \underbrace{\hat{\mathbf{z}}\frac{(M_z - M_0)}{T_1}}_{\text{longitudinal relaxation}} - \underbrace{\frac{(\hat{\mathbf{x}}M_\mathrm{x} + \hat{\mathbf{y}}M_\mathrm{y})}{T_2}}_{\text{transversal relaxation}} . \tag{2.10}$$

The magnetization $\mathbf{M}$ is basically the sum of the expectation values of all participating spins. Thus, this equation describes a macroscopic system, and hence models various quantum mechanical effects. Some of these quantum mechanical effects will be discussed briefly in the subsequent paragraph.

To measure the two different relaxation times independently, CW-ESR is not sufficient. Pulsed measurements are commonly utilized instead. Therefore, the strength and the duration of the perturbation, needed to change the state by $\theta = \pi$ or $\theta = \pi/2$, are measured in a CW experiment. Afterwards, these values can be used to perform advance operation sequences on the spin state. A famous example is the Hahn-Echo [23], which allows the measurement of both relaxation times separately by a $\pi/2$ - π - $\pi/2$ pulse sequence.

Further terms of the spin Hamiltonian

Since the pure spin Zeeman Hamiltonian (eq. 2.2) only represents a free electron spin, the first enhancement is to introduce the bonding of the electron to a nucleus. This is analogous to the model of a standard hydrogen atom Hamiltonian. It yields the orbital momentum and its interaction with the magnetic field. Historically this was called the "normal Zeeman effect", whereas the effect based on the electron's spin was called "anomalous Zeeman effect". From relativistic quantum mechanic it is known that spin and orbital momentum are coupled even though they arise from different quantum mechanical spaces. This is called spin-orbit interaction:

$$\mathcal{H}_\mathrm{EZI} + \mathcal{H}_\mathrm{LS} = \mu_\mathrm{B}\left(\mathbf{L} + g_\mathrm{s}\mathbf{S}\right)\cdot\mathbf{B} + \lambda\mathbf{L}\cdot\mathbf{S}. \tag{2.11}$$

Where $\mathbf{L}$ is the orbital angular momentum and λ the spin-orbit coupling constant. In systems, in which the spin-orbit interaction is strong compared to the external magnetic field, it is convenient to introduce the total angular momentum $\mathbf{J} = \mathbf{L} + \mathbf{S}$. This is valid in free atoms or rare earth atoms which have a high atomic number and the relevant 4f-electrons are located close to the core. In these cases it is common to introduce the Landé g-factor for the total angular momentum [8]:

$$g_{\mathrm{J}} = 1 + \frac{J\,(J+1) + S\,(S+1) - L\,(L+1)}{2J\,(J+1)} \tag{2.12}$$

$$\Longrightarrow \quad \mathcal{H} = g_{\mathrm{J}}\mu_{\mathrm{B}}\mathbf{J}\cdot\mathbf{B}. \tag{2.13}$$

The transformation of the states to this new basis can be done by the application of Clebsch-Gordon coefficients [24].

Moving away from the simple model of a free atom to a specimen in a crystal or molecular environment, the electrostatic potential of the surroundings has to be considered. This additional term is called crystal field and is the leading energy term in most systems (e.g. 3d-transition metals). The main effect of this additional potential is the reduction of the energy degeneracy of the orbital states. Together with the spin orbit coupling this leads to magnetic anisotropy of the crystal. It is common to introduce an effective spin Hamiltonian which combines terms of the crystal field, the spin-orbit coupling and the Zeeman interaction according to their behavior in magnetic resonance experiments:

$$\mathcal{H}_{\mathrm{eff}} = \underbrace{\mu_{\mathrm{B}}\mathbf{B}\cdot\bar{\bar{g}}\cdot\mathbf{S}}_{\text{eff. Zeeman energy}} + \underbrace{\mathbf{S}\cdot\bar{\bar{D}}\cdot\mathbf{S}}_{\text{eff. crystal field}} \,. \tag{2.14}$$

The first term is magnetic-field-dependent and gives rise to similar resonance behavior as before. But it has to be noted that $\bar{\bar{g}}$ is a rank two tensor which is already depending on the crystal field, and therefore renders parts of the anisotropy. The effective crystal field term containing the single ion anisotropy $\bar{\bar{D}}$, in contrast, is magnetic-field-independent and thereby features resonances already at zero field for $S > 1/2$ (zero field splitting). For $S = 1/2$ Kramer's theorem [25] applies, which states that for a system with an odd number of electrons in a purely electric field the ground state is at least doubly degenerate due to time reversal symmetry [26]. The simplest types of the effective crystal field are uniaxial and transversal anisotropy terms. They can be derived by diagonalizing $\bar{\bar{D}}$ and reordering the terms as follows:

$$\mathbf{S}\cdot\bar{\bar{D}}\cdot\mathbf{S} \quad \Longrightarrow \quad D\left(S_{\mathrm{z}} + \frac{1}{3}S\,(S+1)\right) + \frac{1}{2}E\left(S_+^2 + S_-^2\right). \tag{2.15}$$

In this standard form D is the uniaxial anisotropy constant and E the transversal one. $S_\pm$ are the ladder operators, which higher/lower the magnetic quantum number according to $S_\pm\,|S, m_{\mathrm{S}}\rangle = \hbar\sqrt{S\,(S+1) - m_{\mathrm{S}}\,(m_{\mathrm{S}}+1)}\,|S, m_{\mathrm{S}} \pm 1\rangle$. For arbitrary symmetries the effective crystal field is usually expanded into Stevens operators [27], which consist only of polynomial expressions in S_z, $S_\pm$ and constant terms. In this situation it is simple to solve the spin Hamiltonian and calculate the transition rates numerically for a given value of S.

The next interactions to be mentioned are related to the nucleus of the atom. The nucleus itself possesses a magnetic moment $g_\mathrm{I}\mu_\mathrm{n}\mathbf{I}$. There are the hyperfine coupling and the nuclear Zeeman effect which comprise the nuclear magnetic moment, respectively nuclear spin I:

$$\mathcal{H}_\mathrm{HFI} + \mathcal{H}_\mathrm{NZI} = \underbrace{\mathbf{I} \cdot \bar{\bar{A}} \cdot \mathbf{S}}_{\text{hyperfine interaction}} - \underbrace{g_\mathrm{I}\mu_\mathrm{n}\mathbf{B} \cdot \mathbf{I}}_{\text{nuclear Zeeman interaction}} \quad . \tag{2.16}$$

$\bar{\bar{A}}$ is the hyperfine interaction tensor; g_I the corresponding g-factor and μ_n is the nuclear magneton. The hyperfine interaction can be understood analogous to the spin-orbit interaction. In textbooks it is often explained by imagining an additional magnetic field produced by the nuclear core, which acts on the electron spin. Since in most situations the energy splitting of the hyperfine interaction is small compared to the magnetic field, it is not needed to introduce the total angular momentum as a new quantum number. Thus, both m_I and m_S remain good quantum numbers. This is called the Paschen-Back effect, where each resonance is split up into $2I + 1$ lines. The nuclear Zeeman term which renders the interaction of the nuclear moment with the external field is much smaller than all other terms. This is consequential as the nucleus is much heavier than an electron, and hence the nuclear magneton μ_n is three orders of magnitude smaller.

The last term to be mentioned is spin-spin interaction. This is of course a many-particle phenomenon which involves the growth of the Hilbert space according to the number of particles and their degrees of freedom. The most common term is the Heisenberg exchange interaction [28]:

$$\mathcal{H}_\mathrm{ex} + \mathcal{H}_\mathrm{dm} = \underbrace{J_{12}\mathbf{S}_1 \cdot \mathbf{S}_2}_{\text{Heisenberg exchange}} + \underbrace{\mathbf{D}_{12} \cdot (\mathbf{S}_1 \times \mathbf{S}_2)}_{\text{antisymmetric exchange}} \quad . \tag{2.17}$$

It is clear to see that the sign of the coupling constant J_{12} is of great importance. For $J_{12} < 0$ a parallel alignment is energetically preferred, which leads to ferromagnetism. For $J_{12} > 0$ antiparallel alignment is favored, which results in antiferromagnetism. The interaction can be understood by considering two electrons. As they are indistinguishable fermions, their total wave function must be antisymmetric. But the total wave function is a product of spacial and spin component. So, either the spin or the spacial component must be asymmetric.

As there is an energy difference caused by coulomb interaction between the symmetric and the antisymmetric spacial wave function the spin part will render this behavior as well.

The antisymmetric exchange interaction is known as Dzyaloshinsky-Moriya interaction (DMI), after their findings in 1958 [29,30]. This interaction favors a spin canting in magnetically ordered systems, which is comprehensible because of the cross product of the two spins. It is prominent in magnetic skyrmions [31], which is a special magnetic face and will be further explained in the chapter of the sample Fe on Ir(111) (ch. 6).

If the shown spin-spin interactions dominate the system a ferromagnet or antiferromagnet is present. This is covered by ferromagnetic resonance, which will be discussed in the next chapter.

2.1.2 Ferromagnetic resonance

Ferromagnetic resonance is a technique to probe ferromagnetic or antiferromagnetic samples in the same way as in ESR, which was described before. The difference lies in the type of material, its properties and the resulting consequences on the resonance behavior. In ferromagnetics the magnetic moments are not isolated anymore, but strongly interacting via exchange interactions. These interactions are the origin of magnetic order, which itself results in a lager internal magnetic field. It is common in ferromagnetic resonance to introduce the effective field $\mathbf{H}_{\text{eff}} = \mathbf{H}_{\text{int}} + \mathbf{H}_0$ as a composition of internal and external magnetic field (where $\mathbf{H}_0 = {}^{\mathrm{B}_0}/_{\mu_0}$ is the external magnetic field in vacuum). Furthermore, it is sufficient to use a classical description, since common ferromagnetic samples have a vast number of energy levels and the correspondence principle predicts identical results for quantum mechanical or classical derivations [32]. Additionally a homogeneously magnetized sample can be considered which can be described by the macroscopic magnetization $\mathbf{M}$. The equation of motion describing the dynamics of such a sample was proposed 1938 by Landau and Lifschitz [33]. In 1955, Gilbert introduced an improved damping term [34] which yielded the prevalent Landau-Lifschitz-Gilbert equation (LLG):

$$\frac{d\mathbf{M}}{dt} = -\gamma \left(\underbrace{\mathbf{M} \times \mathbf{H}_{\text{eff}}}_{\text{precession term}} - \underbrace{\eta \mathbf{M} \times \frac{d\mathbf{M}}{dt}}_{\text{damping term}} \right). \tag{2.18}$$

This equation of motion was extremely successful in describing dynamics in ferromagnetics over the last decades. Figure 2.3a shows the motion of the magnetization vector $\mathbf{M}$ as described by the LLG equation (eq. 2.18). The magnetization precesses around the effective field $\mathbf{H}_{\text{eff}}$, with a momentum proportional to the gyromagnetic ratio γ and the magnitude of the field (blue arrow). This yields a precession frequency ω_0. The damping (red arrow), with its damping constant η, is perpendicular to $\mathbf{M}$ and $d\mathbf{M}/dt$. As can be seen, the only difference to the Bloch equation (eq. 2.10) is the damping term. This difference is necessary, since the simple Bloch equation cannot explain the lineshapes of resonance experiments in strong fields [35].

To measure the resonance, respectively the precession, in ferromagnetic samples, the same technique as described in the previous chapter (ch. 2.1.1) is used: The introduction of a small oscillating magnetic field can compensate the damping term and therefore excite the system. Figure 2.3b depicts the simplest situation: A rotating field (this can be reformulated through two linear oscillating fields) is applied perpendicular to the effective field $\mathbf{H}_{\text{eff}}$. This additional field produces a torque on the magnetization, which yields an additional term to the LLG equation (2.18):

$$\frac{d\mathbf{M}}{dt} = \ldots - \gamma\mathbf{M} \times \mathbf{H}_{\text{exc}}. \tag{2.19}$$

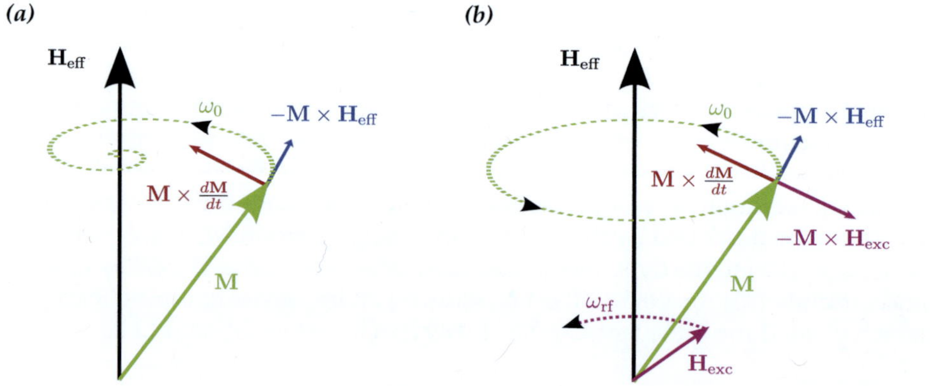

Figure 2.3: ***(a)*** *Sketch of the precession of a magnetization vector* $\mathbf{M}$ *around an effective field* $\mathbf{H}_{\text{eff}}$ *as described by the Landau-Lifschitz-Gilbert equation. The direction of the precession term is shown in blue. The Gilbert damping, which is proportional to the time derivative of the magnetization vector, is depicted in red. The motion of the magnetization vector is a damped precession towards a parallel alignment with the effective field.* ***(b)*** *A small oscillating field* $\mathbf{H}_{\text{exc}}$ *(purple) is applied perpendicular to* $\mathbf{H}_{\text{eff}}$*. If the frequency of this field* ω_{rf} *equals the precession frequency* ω_0 *of the magnetic moment, the torque produced by the oscillating field (*$\mathbf{M} \times \mathbf{H}_{\text{exc}}$*) will continuously compensate the damping.*

If the precession frequency ω_0 equals the oscillation frequency of the additional field ω_{rf}, the torque will continuously compensate the damping. In this resonant situation the energy of the oscillating field is absorbed by the sample. This increase in absorption can be measured in a spectroscopy measurement by sweeping the amplitude of the external field $\mathbf{H}_0$ (which is a component of $\mathbf{H}_{\mathrm{eff}}$) or the frequency ω_{rf}.

The main task for correctly interpreting FMR resonances is to analyze the effective field $\mathbf{H}_{\mathrm{eff}}$. It contains all additional terms and quantum effects that shift the resonance frequency of a ferromagnetic material. In a homogeneous single crystal, upon others, these are the demagnetization field (shape anisotropy) and the magnetocrystalline anisotropy. A common approach is to consider the free energy per unit volume F and the magnetization in spherical coordinates. In the equilibrium state the magnetization and the internal effective field are parallel and the free energy has a minimum:

$$|H_{\mathrm{eff}}| = -\frac{\partial F}{\partial M_s}; \quad \frac{\partial F}{\partial \theta} = 0; \quad \frac{\partial F}{\partial \varphi} = 0. \tag{2.20}$$

With θ and φ being the polar and azimuthal angles of the magnetization and M_{s} the saturation magnetization. Through a linear expansion of the free energy around the equilibrium position (θ_{equ} and φ_{equ}), Suhl [36] solved the undamped equation of motion (eq. 2.18). He found oscillating solutions with the following frequencies:

$$\omega_{\mathrm{res}} = \gamma H_{\mathrm{eff}} = \frac{\gamma}{M_{\mathrm{s}} \sin\theta_{\mathrm{equ}}} \sqrt{\frac{\partial^2 F}{\partial \theta^2}\frac{\partial^2 F}{\partial \varphi^2} - \left(\frac{\partial^2 F}{\partial\theta\partial\varphi}\right)^2}\,\Bigg|_{\theta_{\mathrm{equ}},\varphi_{\mathrm{equ}}} . \tag{2.21}$$

With this solution it is easy to determine the resonance frequency in a ferromagnetic sample. The influence of various additional terms can be determined by extending the free energy with the appropriated terms:

$$F = F_0 + F_{\mathrm{d}} + F_{\mathrm{ua/ca}} + \dots \tag{2.22}$$

Here, F_0 is the interaction with the external field, F_{d} is the demagnetization energy and $F_{\mathrm{ua/ca}}$ is the (uniaxial/cubic) magnetocrystalline anisotropy energy. The influence of these terms on the resonance frequency will be briefly discussed in the following paragraph.

Shape anisotropy

The demagnetization energy reflects the dependency on the sample shape. It is common to consider an ellipsoidal sample, which can be modeled by the introduction of the demagnetization shape factors N_{x}, N_{y} and N_{z} ($N_{\mathrm{x}} + N_{\mathrm{y}} + N_{\mathrm{z}} = 4\pi$). The energy of the demagnetization field F_{d} is given by:

$$F_{\mathrm{d}} = \frac{1}{2}\mu_0 \left(N_{\mathrm{x}} M_{\mathrm{x}}^2 + N_{\mathrm{y}} M_{\mathrm{y}}^2 + N_{\mathrm{z}} M_{\mathrm{z}}^2\right). \tag{2.23}$$

For simplicity it can be chosen that the external magnetic field coincides with the z-axis of the coordinate system. To calculate the resonance frequency with equation 2.21, one has to rewrite the demagnetization energy in polar coordinates and find the equilibrium angles of the magnetization θ_{equ} and φ_{equ}. Differentiation of the energy with respect to the angles then yields:

$$\omega_{\mathrm{res}} = \gamma\sqrt{\left[H_0 + (N_{\mathrm{x}} - N_{\mathrm{z}})\,\mu_0 M_{\mathrm{s}}\right]\left[H_0 + (N_{\mathrm{y}} - N_{\mathrm{z}})\,\mu_0 M_{\mathrm{s}}\right]}. \tag{2.24}$$

This formula was first presented by Kittel in 1948 and is widely known as Kittel equation [37]. For a spherical sample ($N_{\mathrm{x}} = N_{\mathrm{y}} = N_{\mathrm{z}}$) the demagnetization field vanishes and the effective field reduces to the external field. In the case of a thin film perpendicular to the external field $N_{\mathrm{x}} = N_{\mathrm{y}} = 0$ and $N_{\mathrm{z}} = 4\pi$ the effective field will be lowered by a factor of $4\pi\mu_0 M_{\mathrm{s}}$. This means that even at zero external field a nonzero resonance frequency can be found. This is called a zero field gap.

Magnetocrystalline anisotropy

Considering magnetocrystalline anisotropy the easiest form is simple uniaxial anisotropy. The free energy can be expanded as a power series with respect to the sine of the angle between the principle axis and the magnetization vector ϑ:

$$F_{\mathrm{ua}} = K_{\mathrm{u}} \sin^2 \vartheta + \ldots. \tag{2.25}$$

To find the resonance frequency with equation 2.21, different situations have to be considered. Firstly, they depend on the sign of the anisotropy constant K_{u}. And secondly on the orientation of the external field with respect to the principal axis. Let us considers a thin film where the principle axis is perpendicular to the film plane: Positive K_{u} means the minimum energy is given for an out-of-plane

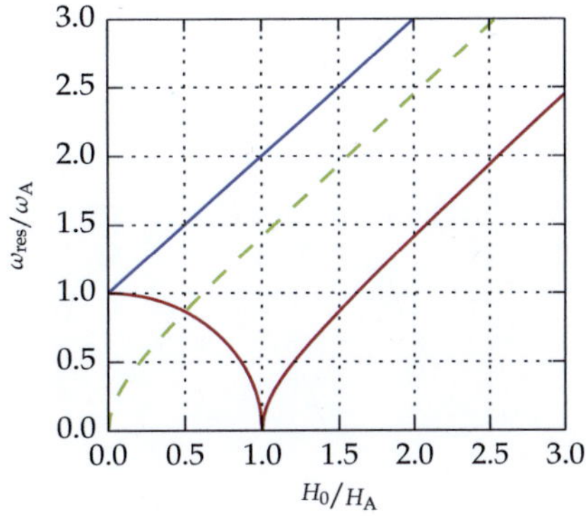

Figure 2.4: *Resonance frequency for a sample with uniaxial anisotropy. The axes are scaled to the anisotropy field $H_A = 2K_u/M_s$, respectively to its resonance frequency ω_A. For the solid lines $K_u > 0$. The blue line shows the resonance for an external field parallel to the principal axis. The red curve is the solution for an in-plane field. The green, dashed line represents a sample with $K_u < 0$ and an in-plane field. For $K_0 < 0$ no resonance can be found for fields out of the thin film plane.*

orientation. The principle axis is thereby the easy axis. In figure 2.4 it can be seen that for a perpendicular orientation of the external field with respect to the film a zero field gap occurs. This means that even at zero external field a resonance can be found at finite frequencies. The frequency of this resonance is directly related to the strength of the anisotropy ($\omega_A = \gamma\,(2K_u/M_s)$). For an external field in the plane of the film a double resonance should occur. This is visible as two solutions can be found for the red curve in figure 2.4 below ω_A. The resonance at $\omega_{res} = 0$ is usually not accessible because even small direction changes of the external field suppress this pole [38].

For materials with negative K_u the ground state is located in the plane of the film. The dashed line in figure 2.4 shows the resonance curve for negative K_u and an external field in the film plane. For these materials no resonance can be found for fields perpendicular to the plane [32].

Since the symmetry of most ferromagnets is cubic, the simple uniaxial anisotropy is insufficient to explain their resonance behavior. The expansion of the free energy in this case is done with respect to the cosines $\alpha_{1,2,3}$ between the magnetization and the cubic axes.

$$F_{ca} = K_1\left(\alpha_1^2\alpha_2^2 + \alpha_2^2\alpha_3^2 + \alpha_3^2\alpha_1^2\right) + \ldots \tag{2.26}$$

Figure 2.5 shows the situations for positive ***(a)*** and negative ***(b)*** anisotropy constant K_1. It can be seen that for a positive anisotropy constant (fig. 2.5***(a)***) the cubic axes are energetically favored (blue). Thereby, they are the easy axes, whereas the space diagonals are the hard axes with the highest energy (red). With a negative anisotropy constant (fig. 2.5***(b)***) the situation is reversed. In both cases the face diagonals are intermediate states. The derivation of the resonance frequencies follows again through equation 2.21. Similar to the uniaxial case zero field gaps and double resonances occur. However, the angle dependence

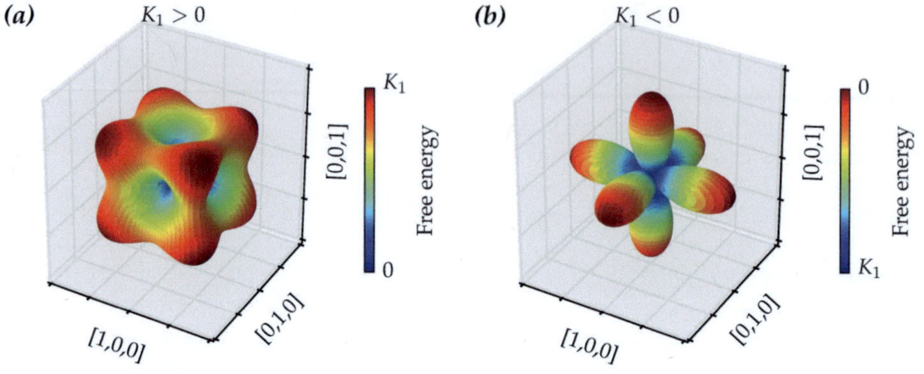

Figure 2.5: *Energy surface for cubic anisotropy for a positive* ***(a)*** *and a negative* ***(b)*** *anisotropy constant K_1. Both color and shape represent the free energy corresponding to different orientations of the magnetization with respect to the crystallographic axes. It can be seen that for $K_1 > 0$ the cubic axes are energetically favored and the space diagonals are the hard axes. For $K_1 < 0$ the situation is reversed.*

between the crystal orientation and the magnetic field is much more complicated in cubic symmetry than in the uniaxial case. A full derivation can be found in [32].

In the previous paragraphs the two examples of shape and magnetocrystalline anisotropy were discussed. They add complexities like zero field gaps, shifts of the resonance point and angular dependencies to the resonance behavior. Additionally, it is noteworthy that the magnetocrystalline anisotropy constants are strongly temperature-dependent. This is logically given when considering a ferromagnet approaching its curie temperature and loosing its ferromagnetism. All these influences must be considered for the design of the experiment, the choice of a sample and the theoretical prediction/verifications of measurements.

2.2 Theory of scanning tunneling microscopy

Historical background

Microscopy, the technique to make small structures visible to the human eye, is a very old scientific field. Since the human eye is sensitive to the visible spectrum of photons, the first microscopes were optical ones. But their resolution is limited by Abbe's diffraction limit $a = {}^{1.22 \cdot \lambda}/_{2 \cdot \mathrm{NA}}$ [39]. Only through "stimulated emission depletion" (STED) and in specific circumstances (fluorescence) this limit was overcome in 1999 by Hell *et al.* [40] (Nobel prize 2014). Of course, other

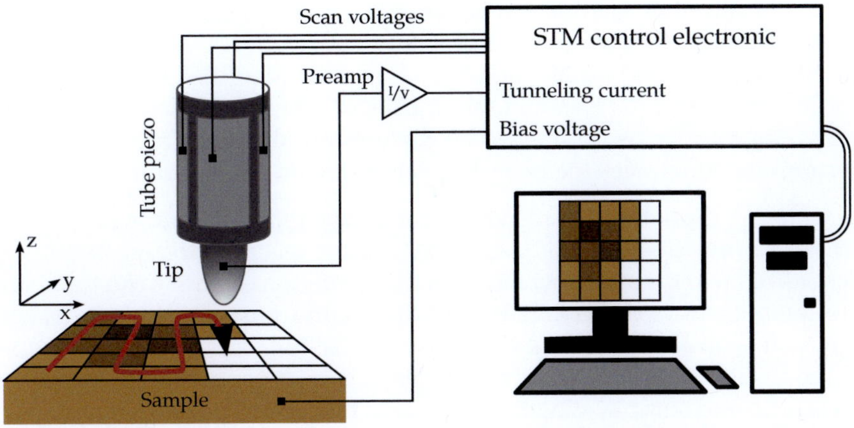

Figure 2.6: Basic principle of an STM. The metallic tip is moved in a grid over the scan area by the use of a piezo-tube. A bias voltage is applied to the sample, which yields a tunneling current flowing from the sample to the tip. This current is amplified by a preamplifier and measured in a control electronic. At every pixel the tunneling current, the position of the tip etc., are recorded by a computer. Thereby, a false color image is produced.

optical microscopes using photons with smaller wavelength were able to access smaller structures much earlier.

To further increase the resolution, electrons with their wave-particle nature were utilized. The resolution is thereby given by the momentum of the electrons, since the de-Broglie relation links it to the wavelength $\lambda = h/p$ [41]. In this way various different kinds of electron microscopes were build, which achieved subatomic resolution. But these microscopes have a few major drawbacks, i.e. the high momentum of the electrons. This can modify or even destroy samples, especially soft ones like organic molecules etc. The even bigger drawback is that in a transmission electron microscope (TEM) the sample has to be partially translucent.

In 1982, the scanning tunneling microscope (STM) was introduced by Binnig and Rohrer [1]. It is capable of accessing surfaces on the atomic scale without the disadvantages mentioned above. Its invention laid to the foundation of the whole field of scanning probe microscopy.

Basic principle

All scanning probe microscopes are based on the principle of placing a probe very close to the sample surface and move it in a grid over the scan area (fig.

2.6). On the atomic scale this can be done using piezoelectric actuators. These materials change their shape on the order of $10\,\mathrm{nm/V}$ once an electrical field is applied. On every grid point (called pixel) certain quantities are measured and recorded by a computer system. By means of a feed back loop the scanning process can directly be influenced by the measured values (see chapter 2.2.2). The computer assembles the recorded data to create a false color image.

In a scanning tunneling microscope a metal tip, typically made of W, is used. Once tip sample distances of less than $1\,\mathrm{nm}$ are reached, the electronic wave functions of tip and sample overlap. On this condition electrons can tunnel from tip to sample and vice versa. This is a pure quantum-mechanical effect, whose origin will be explained in detail in the next section. It is already noteworthy that the resulting tunneling current has a strong dependence on the distance between tip and sample, which can be used to record topography images. As the tunneling current is in the pA to nA range, it needs to be amplified by means of a preamplifier (fig. 2.6). Since even the best available preamplifiers are limited in bandwidth to several kHz, STM is a rather slow technique. For example, if the measurement of every pixel of an image takes about $1\,\mathrm{ms}$, a 512×512 image already takes more $4\,\mathrm{min}$ than to complete. With longer averaging, measurement times of many hours are no exception.

2.2.1 Quantum mechanical tunneling

Electrons, the charge carriers in metals, are quantum mechanical objects. In low energy approximation their stationary states are solutions of the Schrödinger equation:

$$\left[-\frac{\hbar^2}{2m_\mathrm{e}}\nabla^2 + V(\mathbf{r})\right]\psi(\mathbf{r}) = E\psi(\mathbf{r}), \tag{2.27}$$

where $V(\mathbf{r})$ is the potential landscape. The complex solutions of this equation $\psi(\mathbf{r})$ are wave functions with an energy E. The square of their absolute value is interpreted as the probability distribution to find the electron at a given position. It will be shown that electrons, because of their wave nature, can pass a potential barrier which is higher than their kinetic energy. This contrasts with classical objects, which are reflected when their energy is to low. In reference to STM the tunneling barrier represents the gap between tip and sample.

Let us consider a simple, one-dimensional potential barrier with thickness d (fig. 2.7). Outside the barrier the potential shall be zero and inside it shall have a constant value $V = V_0$. For all three regions (I, II and III) the Schrödinger

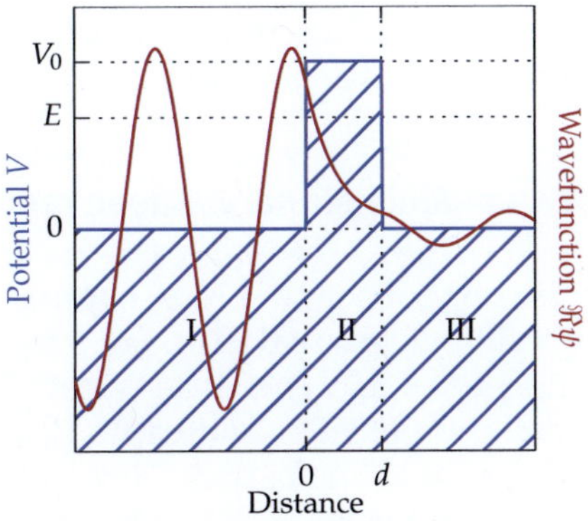

Figure 2.7: Tunneling through a rectangular potential barrier. The blue potential is divided into three constant areas I, II and III. The particle wavefunction is impinging on the wall from the left side. The real part of this wavefunction is drawn in red. In area II (inside the barrier) the exponential decay is noticeable. On the right side, the probability to find the particle, respectively the wave function, is reduced.

equation (2.27) has to be solved separately. It is common to make the following ansatz for each of the three areas:

$$\psi_{\mathrm{I}}(x) = A_1 e^{ik_{\mathrm{x}}x} + A_2 e^{-ik_{\mathrm{x}}x} \tag{2.28a}$$

$$\psi_{\mathrm{II}}(x) = B_1 e^{\kappa_{\mathrm{x}}x} + B_2 e^{-\kappa_{\mathrm{x}}x} \tag{2.28b}$$

$$\psi_{\mathrm{III}}(x) = C_1 e^{ik_{\mathrm{x}}x} + C_2 e^{-ik_{\mathrm{x}}x}. \tag{2.28c}$$

On the left and right hand side of the barrier plane waves with $k_{\mathrm{x}} = \sqrt{2m_{\mathrm{e}}/\hbar E}$ are the solutions. Inside the barrier an exponential decay with $\kappa_{\mathrm{x}} = \sqrt{2m_{\mathrm{e}}/\hbar\,(V_0 - E)}$ is found. The constant A_1 can be arbitrarily set to one. C_2 is set to zero since the electrons are merely expected to be impinging on the barrier from the left side. All other constants are determined by connection conditions between the three areas. The most interesting one is the square of the constant C_1, which can be identified as the transmission coefficient T:

$$T = |C_1|^2 = e^{-2\kappa_{\mathrm{x}}d}. \tag{2.29}$$

This coefficient gives the probability to find the electron on the right side after quantum-mechanical tunneling through the barrier. Figure 2.7 shows the full solution of the one-dimensional tunneling problem. It can be seen that the wave function is non-zero in area III. As the square of the absolute value of the wave function yields the probability to find the electron at a certain position, it can be understood that the electron can traverse the barrier with finite probability. The exponential dependency of the transmission to the barrier width is the essential characteristic for the operation of an STM. In an experiment the transmission can be determined by measuring the current between left and right side. Thereby, a quantity which can be used to determine the distance between the two sides is found. In the experiment the left and right side represent the tip and the sample, whereas the barrier corresponds to the vacuum or air gap in between.

Three-dimensional Bardeen model

The previous model is, of course, way to simple when considering tunneling conditions in a real STM. For a more accurate description a three-dimensional model of a many-particle system has to be used. Tip and sample can be represented as potential landscapes $V_{\mathrm{S/T}}$, originating from the atomic cores, separated by the tunneling barrier. This system can be solved by applying first-order perturbation theory as shown by Bardeen in 1961 [42]. When tip and sample are far away from each other, both systems satisfy their own stationary Schrödinger equations:

$$\left[-\frac{\hbar^2}{2m_\mathrm{e}}\nabla^2 + V_\mathrm{S}\right]\psi_\mu = E_\mu\psi_\mu \qquad \text{(sample)} \tag{2.30a}$$

$$\left[-\frac{\hbar^2}{2m_\mathrm{e}}\nabla^2 + V_\mathrm{T}\right]\chi_\nu = E_\nu\chi_\nu \qquad \text{(tip).} \tag{2.30b}$$

The solutions of the two separate systems are assumed to be known. As soon as both electrodes are in close distance, the wave functions ψ_μ and χ_ν will not evolve accordingly to the equations 2.30a and 2.30b, but will follow the Schrödinger equation of the full system. This means that the states of the separated situations are no longer eigenstates of the Hamiltonian. An initial eigenstate of one of the two electrodes now has a probability to transfer into a state of the other electrode. This probability is dominated by the transfer matrix element ($M_{\mu\nu}$) between the participating states (ψ_μ and χ_ν). Bardeen derived the tunneling matrix element and translated it, using Green's theorem, into a surface integral over the separation surface Σ between tip and sample:

$$M_{\mu\nu} = M^*_{\nu\mu} = \frac{\hbar}{2m_\mathrm{e}}\int_\Sigma (\psi_\mu\nabla\chi^*_\nu - \chi^*_\nu\nabla\psi_\mu)\cdot d\mathbf{S}. \tag{2.31}$$

The evaluation of the tunneling matrix element remains the main task and reflects the whole geometry of a specific problem [43]. To find the tunneling current in Bardeen's model all tunneling channels have to be summed up. In the continuum limit this sum is replaced by an energy integral over the density of states of tip and sample (ρ_T and ρ_S). For finite temperatures the population of fermions, which is given by the Fermi distribution, has to be taken into account.

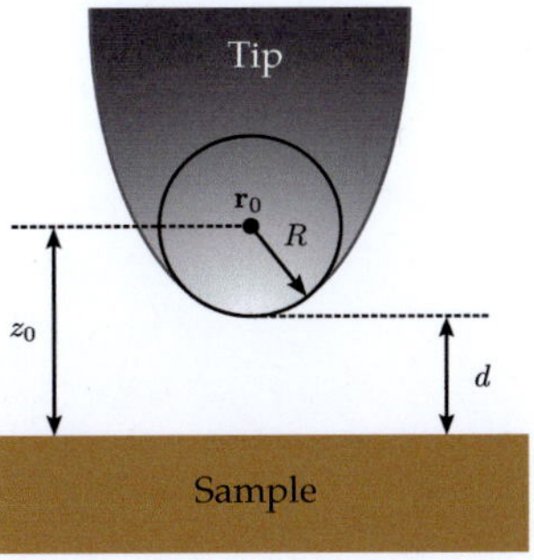

Figure 2.8: Visualization of the Tersoff-Hamann approximation [44]. The tip is modeled as a sphere with radius R. The center of the sphere is located at $\mathbf{r}_0$. In the Tersoff-Hamann model the tunneling current is proportional to the density of states of the sample at this position $\mathbf{r}_0$.

This is done by weighing the density of states accordingly. The tunneling current is then given by:

$$I = \frac{4\pi e}{\hbar} \int_{-\infty}^{\infty} \left[f\left(E_\mathrm{F} - eU + \epsilon\right) - f\left(E_\mathrm{F} + \epsilon\right) \right] \times \rho_\mathrm{T}\left(E_\mathrm{F} - eU + \epsilon\right) \rho_\mathrm{S}\left(E_\mathrm{F} + \epsilon\right) |M_{\mu\nu}|^2 d\epsilon. \quad (2.32)$$

Here, U is the finite bias voltage that lifts the potential of the tip and E_F is the Fermi energy. For temperatures below the desired energy resolution of the experiment, the Fermi distribution can be approximated by the step function. This reduces the integration range to $[0, eU]$. The exponential decay, which was found in the one dimensional case, is hidden in the matrix element as it contains the wave functions with decays into the barrier region.

Approximation by Tersoff & Hamann

Only one year after the introduction of the STM, Tersoff and Hamann applied Bardeen's concept of tunneling to the STM geometry [44,45]. They modeled the tip as a locally spherical potential well with a radius R at a center $\mathbf{r}_0$ (fig. 2.8). They used a s-type wave function for the tip and a two-dimensional Fourier expansion for the sample wavefunction. This allowed them to determine the tunnel matrix element. To find the actual tunneling current, the derivation of Bardeen (eq. 2.32) was used. Energy independent density of states for the tip and operation at low temperatures (density of states should not vary to much within $k_\mathrm{B}T$) were assumed. This yields the central result of Tersoff and Hamann's work:

$$I = \frac{4\pi e}{\hbar} \int_0^{eU} \rho_{\mathrm{T}} \left(E_{\mathrm{F}} - eU + \epsilon\right) \rho_{\mathrm{S}} \left(E_{\mathrm{F}} + \epsilon\right) |M|^2 \, d\epsilon$$

$$\approx \frac{4\pi e}{\hbar} \rho_{\mathrm{T}} \int_0^{eU} \rho_{\mathrm{S}} \left(E_{\mathrm{F}} + \epsilon, \mathbf{r}_0\right) d\epsilon \quad (2.33)$$

$$\rho_{\mathrm{S}} \left(E, \mathbf{r}_0\right) \equiv \left|\psi \left(\mathbf{r}_0\right)\right|^2 \rho_{\mathrm{S}} \left(E\right) \quad (2.34)$$

Here, $\rho_{\mathrm{S}} \left(E, \mathbf{r}_0\right)$ is defined as the *local density of states* (LDOS) of the sample. In this approximation the tunneling current is proportional to the integrated LDOS of the sample at the tip position.

Experimental implementations usually allow direct measurements of the first derivative of the tunneling current with respect to the bias voltage (dI/dU). This tunneling conductance is, in the Tersoff-Hamann approximation, directly proportional to the LDOS of the sample:

$$G\left(U_0\right) \equiv \left.\frac{dI}{dU}\right|_{U=U_0} \propto \rho_{\mathrm{S}} \left(E_{\mathrm{F}} + eU_0, \mathbf{r}_0\right). \quad (2.35)$$

Of course, the limitations of this approximation have to be considered. Tersoff and Hamann estimated the validity of their simple assumption, that the tip only consists of s-type wave functions, to be only true for length scales above $0.3\,\mathrm{nm}$. At smaller sizes wave functions with angular momentum $l \neq 0$ have to be taken into account.

2.2.2 Operation modes of a conventional STM

Following the results of Tersoff and Hamann, an STM is usually operated in two different modes: the constant current and the constant height mode. In constant current mode the current flowing between tip and sample is kept constant by adjusting the distance between tip and sample. This is done by a feedback loop in the control electronic. It changes the voltage on the z-piezo according to the measured current. If the current increases, the distance is increased to keep the current constant. If the current decreases, the tip is moved closer to the sample. When scanning a surface in this mode a topography image containing the integrated density of states of the sample is obtained. This mode is typically used to retrieve the "geometrical" structure of a sample as it is robust against (big) spacial changes.

The opposite is the constant height mode, where the feedback loop is disabled. Thus, the distance between tip and sample is fixed at a certain value. In this case, moving the tip over the sample is dangerous as a big heightening on the sample might lead to contact between the latter and the tip. As a consequence, constant height is mainly used for spectroscopic investigations. They are based on the variation of a specific parameter and the recording of the tunneling current or its derivative. Most prevalent is scanning tunneling spectroscopy (STS), where the bias voltage is swept over a specific range and the IV curve and/or the differential conductance (dI/dU) is recorded. According to equation 2.35, the obtained spectra contains the local density of states at the tip position. Other parameters can be varied as well. In particular the magnetic resonance measurements in an STM, which will be introduced later, could be based on frequency sweeps in constant height mode.

It should be mentioned that both modes can also be combined in one measurement. For example, the tip can be scanned over the surface in constant current mode. Additionally, a change into constant height mode is done at every pixel and a spectroscopic measurement is performed. Afterwards, the scan is continued in constant current mode. For a grid lattice of STS measurements this returns a dI/dU-map. Not only (square) grids are useful but also lines or other distributions of spectroscopy points.

2.2.3 Spin-polarized STM

The idea to measure magnetic information with an STM is almost as old as the STM itself. It reduces to measurement of the spin polarization of a sample and hence the need of a spin polarized tip. Although there were many ideas and approaches to use superconducting [46,47] or semiconducting [48,49] tips, ferromagnetic tips became the quasi standard. The situation in an STM thus corresponds to a ferromagnet-insulator-ferromagnet tunnel junction, which was already studied in 1975 by Julliere [50]. He found that the tunneling conductance depends on the relative orientation of the magnetization of the two ferromagnets. This phenomenon can be explained since each tunneling electron carries a spin; thus the density of states of the ferromagnetic electrodes must be separated into a majority ($\rho_\uparrow$) and a minority ($\rho_\downarrow$) band. The spin polarization of a ferromagnet at the Fermi energy is then defined as follows:

$$P = \frac{\rho_\uparrow - \rho_\downarrow}{\rho_\uparrow + \rho_\downarrow}. \tag{2.36}$$

The influence of spin-polarized electrodes in a tunneling experiment can be understood by extending Bardeen's model of tunneling. This is done by the introduction of two separate spin channels. The single component wave functions of tip and sample are replaced by *spinors*:

$$\mathbf{\Psi} = \begin{pmatrix} \psi_{\uparrow}(\mathbf{r}) \\ \psi_{\downarrow}(\mathbf{r}) \end{pmatrix} e^{-iE_{\mu}/\hbar}. \tag{2.37}$$

Analogous to the spin averaged case, the Pauli equation, which is the Schrödinger equation for spinors, is solved in the tip and the sample area separately. It is simple to define a reference direction where, for example, the tip spinors are diagonal:

$$\mathbf{\Psi}_{\mathrm{T},\uparrow} = \chi(\mathbf{r}) \begin{pmatrix} 1 \\ 0 \end{pmatrix}, \qquad \mathbf{\Psi}_{\mathrm{T},\downarrow} = \chi(\mathbf{r}) \begin{pmatrix} 0 \\ 1 \end{pmatrix} \tag{2.38a}$$

$$\mathbf{\Psi}_{\mathrm{S},\uparrow} = \psi(\mathbf{r}) \begin{pmatrix} \cos(\theta/2) \\ i\sin(\theta/2) \end{pmatrix}, \qquad \mathbf{\Psi}_{\mathrm{S},\downarrow} = \psi(\mathbf{r}) \begin{pmatrix} i\sin(\theta/2) \\ \cos(\theta/2) \end{pmatrix}. \tag{2.38b}$$

With χ and ψ being the wave function as in the non-spin-polarized case. And θ being the angle between tip and sample spin polarization. In this frame the tunneling matrix element, which is now a 2×2 matrix of the different tunneling paths becomes:

$$M_{\mu\nu} \Rightarrow \begin{pmatrix} M_{\uparrow\uparrow} & M_{\uparrow\downarrow} \\ M_{\downarrow\uparrow} & M_{\downarrow\downarrow} \end{pmatrix}_{\mu\nu} = M_{\mu\nu} \cdot \begin{pmatrix} \cos(\theta/2) & i\sin(\theta/2) \\ i\sin(\theta/2) & \cos(\theta/2) \end{pmatrix} \tag{2.39}$$

In the easiest configurations (parallel or antiparallel orientation) this is a diagonal or antidiagonal matrix. The tunneling conductance between to ferromagnetic electrodes is then simply given by:

$$G_{\mathrm{P}} \propto \rho_{\mathrm{I},\uparrow}\rho_{\mathrm{II},\uparrow} + \rho_{\mathrm{I},\downarrow}\rho_{\mathrm{II}\downarrow} \qquad \text{(parallel)} \tag{2.40a}$$

$$G_{\mathrm{AP}} \propto \rho_{\mathrm{I},\uparrow}\rho_{\mathrm{II},\downarrow} + \rho_{\mathrm{I},\uparrow}\rho_{\mathrm{II}\downarrow} \qquad \text{(antiparallel)}. \tag{2.40b}$$

Figure 2.9 visualizes this situation. In parallel configuration (eq. 2.40a) many minority electrons can tunnel into many empty minority states. Additionally, some majority electrons can tunnel into some majority states. This yields a high conductance. In the antiparallel configuration (eq. 2.40b) many minority electrons find only some empty states in the other electrode. And a few majority electrons face many empty states after tunneling. This results in a lower conductance.

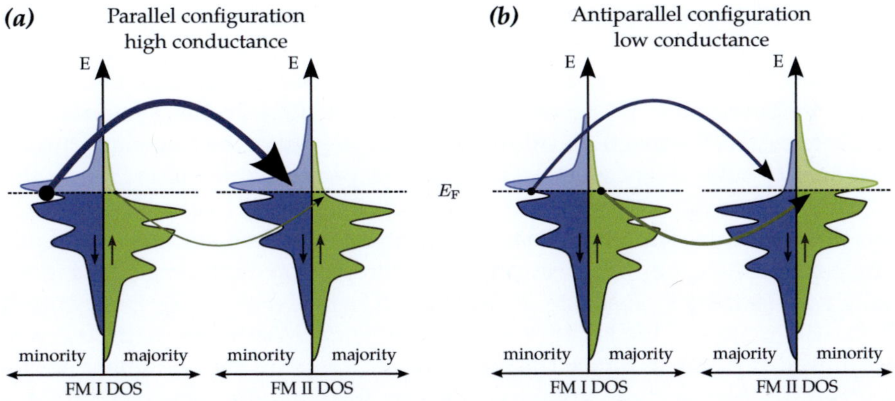

Figure 2.9: *Spin-polarized tunneling between two ferromagnetic electrodes I and II. The densities of states of the electrodes are split into up and down parts, respectively a minority band and a majority band. Tunneling between the electrodes, in absence of spin flip scattering, can only happen in the paths indicated by the arrows. The thickness of the arrows indicates tunneling probability.*

In a non collinear state the full conductance can be written as:

$$G = 2\pi G_0 \left|M_{\mu\nu}\right|^2 (\rho_T \rho_S + m_T m_S \cos\theta) = 2\pi G_0 \left|M_{\mu\nu}\right|^2 \rho_T \rho_S (1 + P_T P_S \cos\theta). \quad (2.41)$$

Here, the spin-polarized $m = \rho_\uparrow - \rho_\downarrow$ and the spin averaged $\rho = \rho_\uparrow + \rho_\downarrow$ part of the density of states were used. They can be transcribed using the tip and sample spin polarizations P_T and P_S. This angular dependency was experimentally verified many times, e.g. by Miyazaki and Tezuka in 1995 [51]. The relative variation of the tunneling conductance between the parallel and the antiparallel state is called junction magnetoresistance or *tunnel magnetoresitance* (TMR):

$$TMR \equiv \frac{G_{max} - G_{min}}{G_{max}} = \frac{2 P_T P_S}{1 + P_T P_S}. \quad (2.42)$$

This ratio reflects the strength of the spin-polarized signal; therefore, it is consequential to maximize it, that is using electrodes with large spin polarizations.

To utilize spin-polarized tunneling in an STM (SP-STM), it is crucial to have a spin-polarized tip. This is usually achieved through coating of a non-magnetic tip with ferromagnetic or antiferromagnetic material. Furthermore, it is important to separate the spin-polarized part of the conductance from the non spin-polarized-one. The first SP-STM experiments were conducted in constant current mode with a CrO_2 tip on a Cr(001) sample [52]. In this system it was possible to extract the magnetic information directly out of the tunneling current. But almost all subsequent works were done in spectroscopic mode [53]. In this mode the differential conductance is measured at energies of spin-polarized states of the sample. This yields magnetic contrast if the non-spin-polarized part of the conductance, i.e. the electronic structure, is constant over the scanned area. By measuring dI/dU-maps, as explained before (ch. 2.2.2), it is possible to produce images of the lateral magnetic structure. With this method it was, for example, possible to reveal the magnetic structure inside a magnetic vortex core [54,55], which will be explained later (ch. 5). A good review of SP-STM of magnetic structures and thin films can be found in reference [6].

3 Magnetic resonance in scanning tunneling microscopy

In the previous chapter the two independent research methods, magnetic resonance and scanning tunneling microscopy were explained. It was shown that magnetic resonance is capable of measuring the dynamic behavior of spins, respectively magnetic moments down to the ns or ps scale. The sensitivity of the experiment limits the minimum amount of spin centers per sample to about 10^{10} [56]. Only with high efforts (e.g. high field, low temperatures, etc.) this detection limit could be decreased to 10^8 spins [57].

Alternatively, STM is an excellent tool to probe single atoms or molecules. Its lateral resolution down to the Å range allows identification, structural [43] and magnetic [58,59] characterization of such nano-objects. A massive limitation is caused by the time resolution at which the dynamics of magnetic objects can be captured in an STM. This arises from limited bandwidth of the transimpedance amplifier, which is used to convert the nA current in an STM to a manageable voltage (ch. 4.1.5). Typical bandwidths are around 10 kHz, which is way to slow when tackling dynamics of magnetic objects. As mentioned above, the time scales of nano-scale magnetic objects strongly depend on their size. The magnetic moment of most atoms or molecules are stable for much less than 1 µs, which is way to short to be detected through the transimpedance amplifier [60]. Only indirect determination of the lifetime (through analysis of the excitation linewidth) conveys an idea of the dynamic motion of the magnetic moment (e.g. the precession). Merely by symmetry arguments [12] or by coupling of many moments to clusters [61] some well chosen examples were found that exhibit large enough lifetimes to be detected directly with a transimpedance amplifier.

The combination of magnetic resonance and scanning tunneling microscopy is meant to overcome this bandwidth limitation. It should allow to measure the dynamics of single magnetic moments in the nanosecond and nanometer range.

In the following sections a few previous approaches with more or less successful attempts of measuring magnetic dynamics in an STM will be reviewed. Through these experiments it will be clear which problems need to be solved. The new concept, which is based on a heterodyne detection of resonance signals in an STM, will be explained afterwards. It is the main idea behind this work. At the end of the chapter a closely related experiment which implements the new measurement concept in a solid state device will be shown.

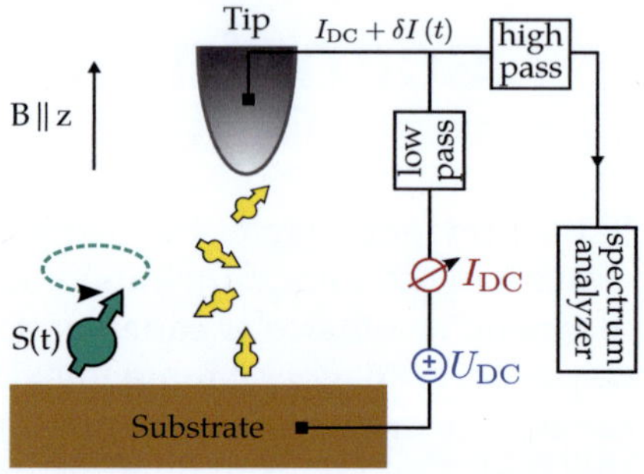

Figure 3.1: *Schematic view of an ESN-STM experiment according to [62,63]. The localized spin $S(t)$ precesses around the magnetic field. At that it modulates the tunneling electrons (yellow) via exchange interaction. The tunneling current is split into a constant and an oscillating part. The oscillating part is detected by means of a spectrum analyzer.*

3.1 Previous attempts and related experiments

3.1.1 Electron-spin-noise STM

The first reports on measuring high-frequency signals in an STM were published in 1989 by Manassen and coworkers [64]. They claim to have detected a radio frequency modulation of the tunneling current originating from an individual spin on a partially oxidized Si(111) surface. To split the spin states, they mounted their sample on a permanent magnet. With this procedure it is impossible to vary the magnetic field in situ for verification of the emerging resonance peaks. The detection was done by separating the high-frequency part of the tunneling current by a high-pass - low-pass network and detecting it by means of a spectrum analyzer (fig. 3.1). In 1993, they established this technique as electron-spin-resonance STM (ESR-STM) [65], which was later rephrased to electron-spin-noise-STM (ESN-STM) [66]. The physical principle of how the local precessing moment can modulate the tunneling barrier was explained by direct exchange coupling between the local moment and the tunneling electrons [62]. A full quantum mechanical description, using second quantization Green's functions, was given by Fransson *et al.* [67]. This technique was also applied to detect the electron spin of typical ESR molecules containing a free radical (BDPA, TEMPO or DPPH) [66,68]. Although there were some massive improvements over the last two decades in this kind of experiments, like advanced electronics or usage of an electro magnet for in situ variations of the electric field, they still lack high reproducibility and good signal-to-noise ratio. In some of the presented measurements only 0.5 % of the spectra show a single-data-point spike above the noise background [66](fig. 3.2). This is understandable since this kind of experiments do not have an active excitation of the system. This would preserve a continuous precession of the spin system. When the relaxation times are way below the sweep rate of the utilized spectrum analyzer, only fractions of the signal are detected. For advancement in the field of ESN-STM the authors state that measurements of noise fluctuations on the time scale at which they are created is inevitable.

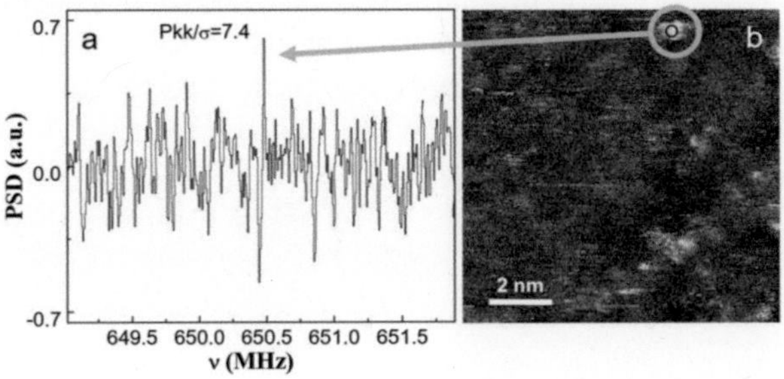

Figure 3.2: *ESN-STM measurement of DPPH molecules on Au(111). On top of the molecule in* ***(b)*** *a spectrum shown in* ***(a)*** *could be recorded showing a peak at* 651.5 MHz. *Reprinted with permission from [66]. Copyright [2007], AIP Publishing LLC.*

3.1.2 Pump probe measurements

An alternative way to access small timescales is the use of pump-probe techniques, which are common in optics. They rely on the nonlinearity of the system under study, as can be seen in figure 3.3a. In an STM this is fulfilled as the I-V characteristic of a tunnel junction is non-linear. Two voltage pluses of different heights (pump- and probe-pulse) excite the system. Because of the nonlinearity there is a difference in current if the pulses arrive simultaneously or separated by a time Δt. Since the signal of just one pump-probe-pulse pair would be too small and too fast, a whole pulse train is send to the system (many thousands of pulse pairs). For reference, a pulse train with $\Delta t \to \infty$ is send afterwards. The statistical response of the system to the pulse sequence is then gained by comparing the signals of the two pulse trains. Nunes, Freeman and parallel Weiss *et al.* were the first to show this kind of experiment in an STM by combining the latter with an optical setup [69,70].

After the development of advanced electrical pulse generators, Loth *et al.* published a more sophisticated, solely electronic-based approach of pump-probe measurement in an STM in 2010 [71]. The setup is depicted in figure 3.3b. They studied the lifetimes of Fe-Cu dimers by measuring their magnetization using a spin-polarized tip. The dimers exhibit two degenerate magnetic ground states, separated by an anisotropy barrier of a certain height. To split these ground states, they applied a magnetic field, thereby favoring one of the states and decreasing the transition rates between them. A voltage pulse higher than the anisotropy barrier (pump-pulse) can inelastically excite the dimer to overcome this anisotropy barrier and reach the opposite magnetic state. The dimer will now fall back into the lowest magnetic state according to its relaxation time T_1 (statistical process). The probe-pulse "reads out" the state of the dimer after the delay time Δt. The amplitude of the probe-pulse has to be lower than the energy of the first excited state to not pump the system again. Depending on the state of the dimer during the probe pulse, a change in current due to spin-polarized tunneling is present. Averaged over a whole pulse train, this yields the statistical decay of the excited state of the dimer. Figure 3.4 shows the different situations during a measurement according to [71]. Using this method it was possible to directly measure spin relaxation time between 50 and 250 ns with atomic resolution.

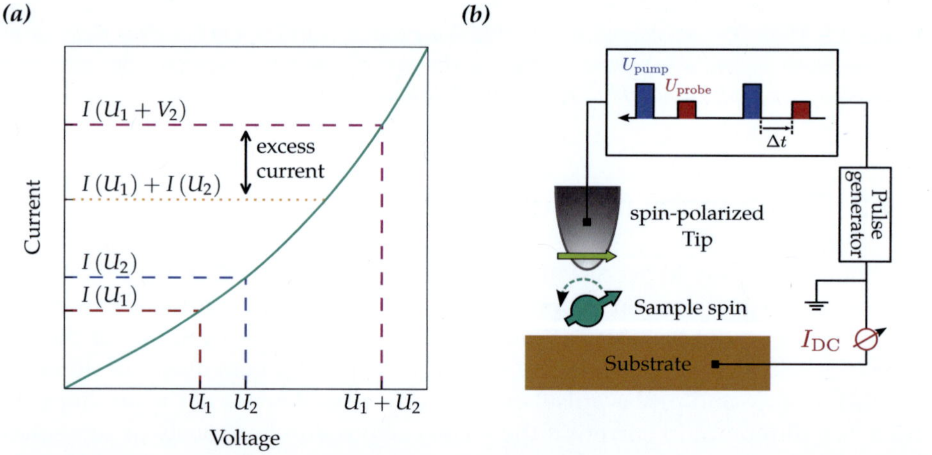

***Figure 3.3:** **(a)**: Visualization of the current excess in a nonlinear tunneling junction according to [69]. The sum of both voltages produces a higher current than the sum of the two separate currents. **(b)**: Schematic setup of a pump probe experiment according to [71]. The pulse trains produced by a pulse generator are send to the STM tip. The excitation through the pump pulse is detected by a spin-polarized tip during the probe pulse. By averaging over a long pulse train the dynamical evolution can be revealed.*

A big issue in this kind of experiments is the pulse shape and its corruption due to transmission inhomogeneity into the tunneling junction of an STM. If an ideal square shape pulse is considered basic math taught that it will contain all

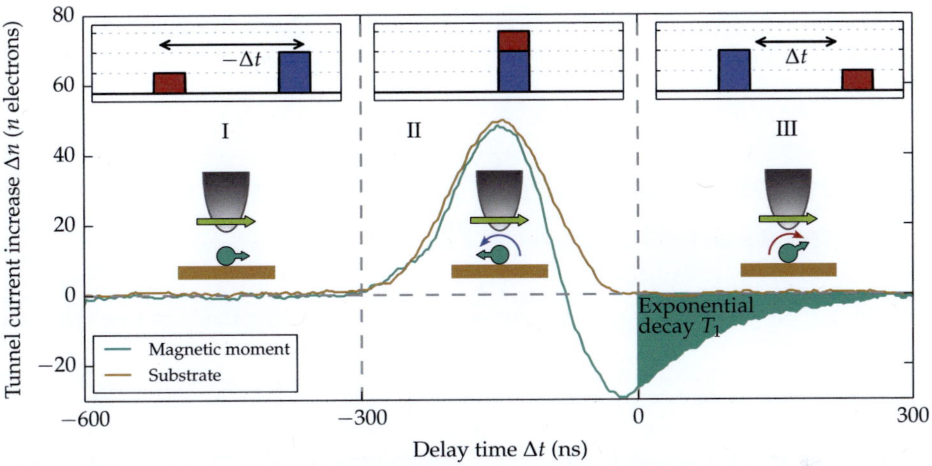

Figure 3.4: *Pump probe experiment adapted from [71]. The upper insets show the real time voltage. The pump pulse (blue) and the probe pulse (red) are separated by a delay time* Δt*. This time is swept through the different areas* I,II *and* III*. In area* I *the magnetic moment is in its ground state when the probe pulse arrives. In area* II *both pulses overlay and an extra current, due to the nonlinearity, is generated. Area* III *shows the situation when the probe pulse arrives after the pump pulse; hence the magnetic moment is still in its excited state. Thus, the spin polarized conductance is decreased due to antiparallel alignment (ch. 2.2.3).*

frequency components up to infinity. Furthermore, it will be shown in chapter 4.2.1 that the transmission of electromagnetic waves inside a cable is frequency-dependent. Therefore, some parts of the pulse will be transmitted differently than others; hence the shape of the pulse will change. To compensate for that, Loth *et al.* used an advanced pulse generator to produce arbitrary pulse shapes. This enabled them to generate modified pulses, so that the transmitted pulses at the tunneling junction met their design pulse shape. To gain knowledge about the actual shape of the pulse at the junction, they made use of the cross correlation of both pulses:

$$R_{xy}(\tau) = (x \star y)(\tau) = \int_{-\infty}^{\infty} x^*(t)\, y(t+\tau)\, dt \tag{3.1}$$

With x being the pump-pulse shape and y the probe-pulse shape. In the tunneling junction both pulses will be mixed and the current will follow the non-linear I-V-curve. This mixing inside the junction exactly corresponds to the calculation of the cross-correlation of the two pulse shapes. It is, therefore, possible to calculate the form of the current response for given pulse shapes. With this utility it was possible to adjust the original pulses in a way to achieve the desired pulse shape at the tunneling junction. The reference spectrum, on the substrate, in figure 3.4 (brown curve) is a nice example of a cross-correlation with well adjusted pulse shapes.

3.2 Heterodyne method for resonance measurements in STM

In this chapter the new measurement concept, which is the basis of this work, is explained.

As was shown in the two previous chapters, it is possible to access the decay time T_1, for timescales on the order of ns, in pump probe experiments. But a measurement of the real spin precession which would allow the process towards full control of the magnetic moment, including measurement of spin-echo, Ramsey frequency etc., similar to conventional ESR, is limited because of two main reasons:

- The detection of radio frequency currents in the nA-range in an STM is still not possible with satisfying performance.
- A continuous excitation, to maintain coherent oscillation over the measurement time, is needed to achieve strong, authentic signals.

It will be shown that the present method is capable of solving both issues at once. They will be addressed separately in the following two sections. The presented technique is based on spin-polarized tunneling and the introduction of a radio frequency bias voltage.

3.2.1 Heterodyne detection of resonance signals using SP-STM

Lets consider an oscillating magnetic moment, e.g. precessing at a frequency ω_0. The spin-polarized conductance, which was presented in chapter 2.2.3, will then vary with this frequency:

$$G(t) = G_{\mathrm{SA}}\,(1 + \underbrace{P_{\mathrm{T}} P_{\mathrm{S}} A}_{\Lambda} \cos(\omega_0 t)) \,. \tag{3.2}$$

Here, G_{SA} is the spin-averaged part of the conductance. The amplitude factor A accounts for oscillations smaller than full $180°$ rotations. Where 0 means no oscillation and 1 stands for a modulation between totally parallel and totally antiparallel alignment. The factor Λ is an abbreviation for the amplitude of the oscillating part of the conductance. The shown conductance is obvious for macroscopic magnetic moments but also true for localized quantum-mechanical spins [67]. Of course, the relative directions of the polarizations of tip and sample are crucial. Only if the projection of the magnetic moment on the polarization direction of the tip contains an oscillating component, the conductance will vary with this frequency according to equation 3.2.

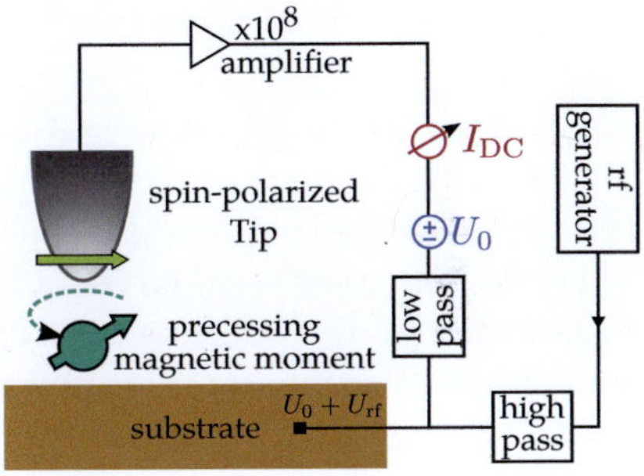

Figure 3.5: *Sketch of an heterodyne magnetic resonance experiment in an STM. Spin-polarized tunneling conditions are present. An oscillating voltage is produced by a rf-generator. This voltage is mixed by means of a high-pass - low-pass network to the bias voltage U_0. High-frequency components will be lost by the amplifier, whereas the heterodyne term can be detected.*

In this spin-polarized tunneling condition, the idea is to apply a radio-frequency (rf) voltage, respectively irradiating microwaves:

$$U_{\mathrm{bias}}(t) = U_0 + U_{\mathrm{rf}} \cos(\omega_{\mathrm{rf}} t + \varphi), \tag{3.3}$$

where U_0 is the bias voltage, U_{rf} the amplitude of the radio frequency voltage and ω_{rf} its angular frequency. The phase φ accounts for the different phases of excitation and precession as will be seen later. Figure 3.5 shows the schematic setup with a spin-polarized tip above a precessing magnetic moment. A rf-generator is connected via a high-pass - low-pass network suppling a radio frequency bias voltage. For simplicity a linear IV-curve can be assumed, where the tunneling current can be obtained by multiplication of conductance times voltage:

$$\begin{aligned} I(t) = G(t) \cdot U_{\mathrm{bias}}(t) = &\underbrace{G_{\mathrm{SA}} U_0}_{\text{constant term } := I_0} + \\ &\underbrace{G_{\mathrm{SA}} U_{rf} \cos(\omega_{\mathrm{rf}} t + \varphi) + G_0 U_0 \Lambda \cos(\omega_0 t)}_{\text{radio frequency term } := I_{\mathrm{rf}}} + \\ &\underbrace{G_{\mathrm{SA}} U_{\mathrm{rf}} \Lambda \cos(\omega_0 t) \cos(\omega_{\mathrm{rf}} t + \varphi)}_{\text{heterodyne term } := I_{\mathrm{het}}}. \end{aligned} \tag{3.4}$$

The *constant term* (I_0) is independent of the magnetic moment. It is given by the theory of Tersoff and Hamann as describe in the previous chapter 2.2.1. The *radio frequency term* (I_{rf}) consists only of components oscillating at frequencies ω_{rf} and ω_0. They are both expected, respectively set, in the rf-range. Thereby they are well above the cut-off frequency of the current amplifier and will not be detectable. Surveying the third *heterodyne term* (I_{het}), it can be observed that the product of two trigonometric functions appears. This can be transcribed using the trigonometric identity:

$$\cos(\alpha)\cos(\beta) = \frac{1}{2}\cos(\alpha+\beta) + \frac{1}{2}\cos(\alpha-\beta)\,. \tag{3.5}$$

Application of this identity to the third term in equation 3.4, obviously yields a term with the sum of the two high frequencies ($\omega_{\text{rf}} + \omega_0$) and one with the difference of the frequencies ($\omega_{\text{rf}} - \omega_0$). This mixing of two initial frequencies to create new ones is called heterodyning. Now two different scenarios should be considered: the off-resonance case ($\omega_{\text{rf}} \neq \omega_0$) and the resonance case, where the frequency of the rf voltage equals the oscillation frequency of the magnetic moment ($\omega_{\text{rf}} = \omega_0$). Lets first consider the off-resonance case:

- $\omega_{\text{rf}} \neq \omega_0$

$$I_{\text{het}}(t) \propto \xcancel{\frac{1}{2}\cos\left((\omega_{\text{rf}} + \omega_0)\,t + \varphi\right)} + \xcancel{\frac{1}{2}\cos\left((\omega_{\text{rf}} - \omega_0)\,t - \varphi\right)}. \tag{3.6}$$

Here, both terms will still be time-dependent. The first term's frequency ($\omega_{\text{rf}}+\omega_0$) will be well above the cut of frequency of the current amplifier and therefore vanishes. The second term oscillates at $\omega_{\text{rf}} - \omega_0$. In the off-resonance case this frequency will easily be on the order of 10 kHz to 1 MHz because the magnetic moment oscillates in the MHz to GHz range. Thus, this term will also not be detected by the current amplifier and the whole heterodyne current will vanish. Approaching the resonance case, this will, of course, not be true anymore:

- $\omega_{\text{rf}} = \omega_0$

$$I_{\text{het}}(t) \propto \xcancel{\frac{1}{2}\cos\left((\omega_{\text{rf}} + \omega_0)\,t + (\varphi + \phi)\right)} + \underbrace{\frac{1}{2}\cos\left(\xcancel{(\omega_{\text{rf}} - \omega_0)\,t} - \varphi\right)}_{\text{Time independant term}}. \tag{3.7}$$

Identical to the off-resonance case, the first term will vanish. But the frequency of the second term will decrease until a time-independent component is obtained at $\omega_{\text{rf}} = \omega_0$. This time-independent term obviously depends on the phase difference between the oscillation of the magnetic moment and the rf voltage.

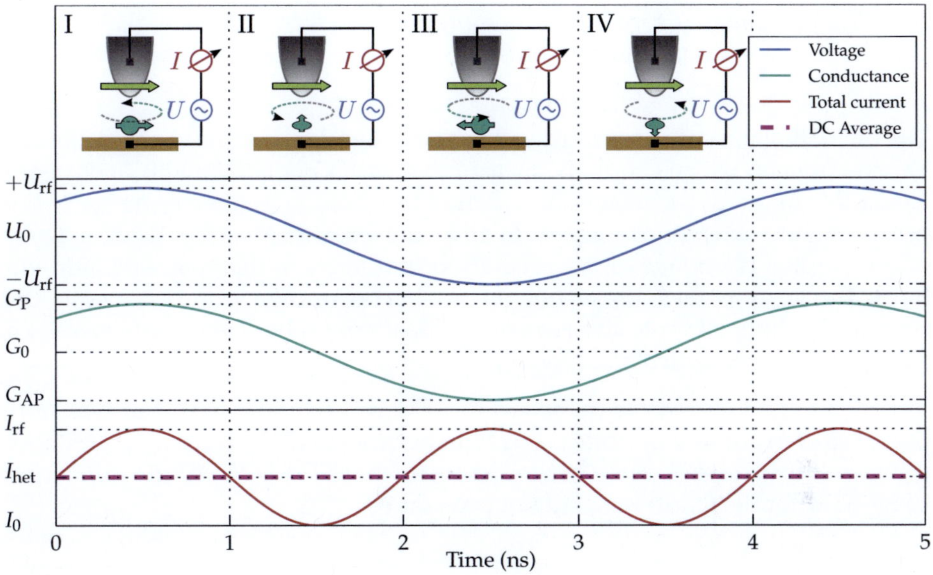

Figure 3.6: *Time sequence visualization of a heterodyne resonance experiment at an arbitrary resonance frequency. The upper line shows a sketch of the orientations between tip and sample magnetization. This creates an oscillating conductance (green curve), between parallel (G_P) and antiparallel (G_{AP}). Along with the radio frequency voltage (blue) this yields the red current. The low-pass behavior of the I-V converter will measure the purple DC component. An in-phase relation between voltage and conductance is depicted.*

For phase differences $\varphi \neq n\pi + \pi/2$ ($n \in \mathbb{N}$) this term will yield a DC component in the tunneling current. This component can be detected respectively processed by the current amplifier.

Figure 3.6 shows a time line of the resonant case in a measurement as described above. The voltage and conductance are oscillating in-phase which produces a DC component in the current:

I: The rf-voltage is positive. The tunneling junction is in parallel configuration, which has high conductance. A high additional current will flow.

II: The rf-voltage is zero. The tunneling junction is in perpendicular configuration, which has intermediate conductance. No additional current will flow.

III: The rf-voltage is negative. The tunneling junction is in antiparallel configuration, which has low conductance. A high additional current will flow.

IV: The rf-voltage is zero. The tunneling junction is in perpendicular configuration, which has intermediate conductance. No additional current will flow.

The oscillating part of the current will not be detectable, but the average yields a DC component, which should be measurable with a conventional I-V converter. Using this heterodyne detection, it should be possible to directly measure the precession of a magnetic moment. In an actual experiment either the frequency of the applied rf-voltage or the oscillation frequency of the magnetic moment has to be swept to reach the resonant case. In cooperation with J. Fransson, the previously shown simple current voltage description has been confirmed by a full quantum mechanical analysis [72].

The idea of an heterodyne detection in STM is not totally new, but was only considered in combined optical - STM experiments [73]. Recently a Japanese group published results on electronic heterodyne STM [74], but the utilization of SP-STM for detection was still not considered.

Advanced signal detection using modulation of the radio frequency

In conventional scanning tunneling spectroscopy it is common to detect the variation of the tunneling current by means of a Lock-in amplifier. The advantage is that omnipresent noise in the tunneling current can be clipped very efficient. In the present heterodyne case this is also helpful since the amplitude of the actual signal is supposed to be small compared to external influences. To make use of a Lock-in amplifier, the measured quantity needs to be modulated at a frequency ω_{mod}. This can be pictured as turning the heterodyne resonance signal on and off periodically. Two obvious methods should be mentioned and are shown in figure 3.7:

- Periodically altering the amplitude of the radio frequency voltage. This is called amplitude modulation (AM). Figure 3.7 red curve.
- Sweeping the frequency periodically in and out of resonance. This is called frequency modulation (FM). Figure 3.7 green curve.

As mainly AM was used in the experiments, the subsequent considerations will be related to this type of modulation. It coincides with the following substitution:

$$U_{\mathrm{bias}} \quad \Rightarrow \quad U\left(t\right) = U_0 + U_{\mathrm{rf}}\left[1 + \cos\left(\omega_{\mathrm{mod}} t\right)\right] \cos\left(\omega_{\mathrm{rf}} t + \varphi\right). \tag{3.8}$$

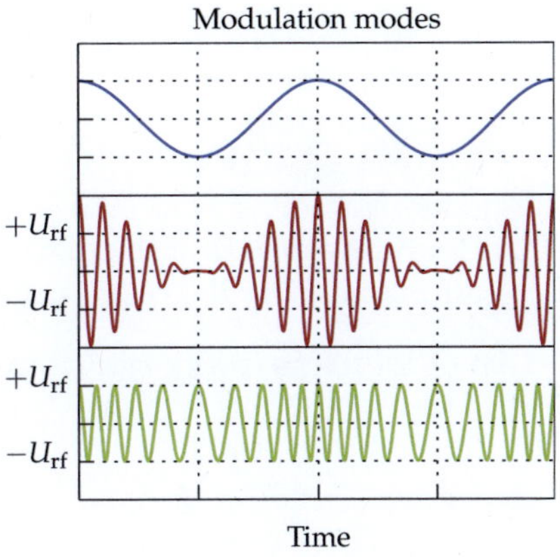

Figure 3.7: *Modulation in a heterodyne resonance experiment. For visibility close frequencies were chosen for the plot. The blue curve shows the modulation frequency (ω_{mod}) at which the Lock-in gathers the signal. The red curve shows 100 % amplitude modulation (AM) of a carrier wave. As can be seen, the amplitude in this case varies between zero and twice the original amplitude. In the lower part, frequency modulation (FM) is plotted in green for completeness.*

With ω_{rf} being the radio frequency in the MHz to GHz range and ω_{mod} the modulation frequency in the kHz regime (below the cutoff frequency of the current amplifier). This voltage formula corresponds to 100 % AM, as depicted in figure 3.7. For the Lock-in amplifier the modulation frequency is the important parameter as it will detect the component of the tunneling current proportional to ω_{mod} (or higher harmonics of ω_{mod}, depending on the settings). It should be noted that AM is already a kind of heterodyne, as the mixing of the two frequencies ω_{mod} and ω_{rf} again yields sidebands at $\omega_{\text{rf}} + \omega_{\text{mod}}$ and $\omega_{\text{rf}} - \omega_{\text{mod}}$. But this time the frequencies are set to be much different; thus, no direct current is created.

To understand the actual signal measured by the Lock-in, it is common to expand the nonlinear, unknown IV-curve in a Taylor series. The expansion is performed up to second order around the bias voltage U_0 because the modulation amplitude U_{rf} is supposed to be much smaller than U_0:

$$I(U,t) \approx I(U_0,t) + \left.\frac{\partial I}{\partial U}\right|_{U_0} (U - U_0) + \frac{1}{2}\left.\frac{\partial^2 I}{\partial U^2}\right|_{U_0} (U-U_0)^2 + \dots \tag{3.9}$$

$$\Rightarrow \quad I(U,t) \approx \underbrace{I(U_0,t)}_{:=I_{0th}} + \underbrace{\frac{\partial I(U_0,t)}{\partial U}\left[U_{\text{rf}}\left(1+\cos\left(\omega_{\text{mod}}t\right)\right)\cos\left(\omega_{\text{rf}}t+\varphi\right)\right]}_{:=I_{1st}}$$
$$+ \underbrace{\frac{1}{2}\frac{\partial^2 I(U_0,t)}{\partial U^2}\left[U_{\text{rf}}\left(1+\cos\left(\omega_{\text{mod}}t\right)\right)\cos\left(\omega_{\text{rf}}t+\varphi\right)\right]^2}_{:=I_{2nd}}. \tag{3.10}$$

Just like in the deduction without modulation, the conductance is expected to have a time-dependent term proportional to $\Lambda\cos(\omega_0 t)$ originating from the oscillating magnetic moment. If the oscillation frequency is bias-independent, it can be extracted from the derivative and again the trigonometric product to sum identity (eq. 3.5) can be used. The component measured by the Lock-in is then found by ordering the expansion and selecting only terms oscillating at ω_{mod} (for simplicity the phase difference φ is neglected in this step):

- 1st order terms (proportional to $\partial I/\partial U$):

$$I_{\mathrm{1st}} = U_{\mathrm{rf}}\frac{\partial I(U_0)}{\partial U}\left\{\cos(\omega_{\mathrm{rf}}t) + \Lambda\cos(\omega_0 t)\cos(\omega_{\mathrm{rf}}t) + \right.$$
$$\left.\cos(\omega_{\mathrm{mod}}t)\left[\cos(\omega_{\mathrm{rf}}t) + \frac{1}{2}\Lambda\cos((\omega_0+\omega_{\mathrm{rf}})t) + \frac{1}{2}\Lambda\cos((\omega_0-\omega_{\mathrm{rf}})t)\right]\right\}. \tag{3.11}$$

- 2nd order terms (proportional to $\partial^2 I/\partial U^2$):

$$I_{\mathrm{2nd}} = \frac{1}{2}U_{\mathrm{rf}}^2\frac{\partial^2 I(U_0)}{\partial U^2}\left\{\cos(\omega_{\mathrm{rf}}t)^2 + \Lambda\cos(\omega_0 t)\cos(\omega_{\mathrm{rf}}t)^2 + \right.$$
$$\cos(\omega_{\mathrm{mod}}t)\left[1 + \cos(2\omega_{\mathrm{rf}}t) + 2\Lambda\cos(\omega_0 t)\cos(\omega_{\mathrm{rf}}t)^2\right] +$$
$$\left.\cos(\omega_{\mathrm{mod}}t)^2\left[\cos(\omega_{\mathrm{rf}}t)^2 + \Lambda\cos(\omega_0 t)\cos(\omega_{\mathrm{rf}}t)^2\right]\right\}. \tag{3.12}$$

Most of these terms will again be cut by the IV-converter, and therefore also not be detected by the Lock-in. As the Lock-in will measure the part of the tunneling current proportional to ω_{mod} (marked in red), the important remaining terms are given by:

$$I_{\mathrm{Lock-in}} = \underbrace{\frac{1}{2}U_{\mathrm{rf}}\frac{\partial I(U_0)}{\partial U}P_{\mathrm{T}}P_{\mathrm{S}}A\cos((\omega_0-\omega_{\mathrm{rf}})t-\varphi)}_{\text{resonance term}} + \underbrace{\frac{1}{2}U_{\mathrm{rf}}^2\frac{\partial^2 I(U_0)}{\partial U^2}}_{\text{static term}}. \tag{3.13}$$

This is the full signal expected in an heterodyne SP-STM measurement. As intended, the time dependence vanishes as soon as a resonance between the external radio frequency and the internal oscillation occurs. The resonance signal should scale with the radio frequency amplitude U_{rf}, the conductance $\partial I(U_0)/\partial U$ and the polarization of tip and sample P_{T} and P_{S}. A non-vanishing polarization of both electrodes and a non-zero oscillation amplitude A are obviously crucial to obtain a signal.

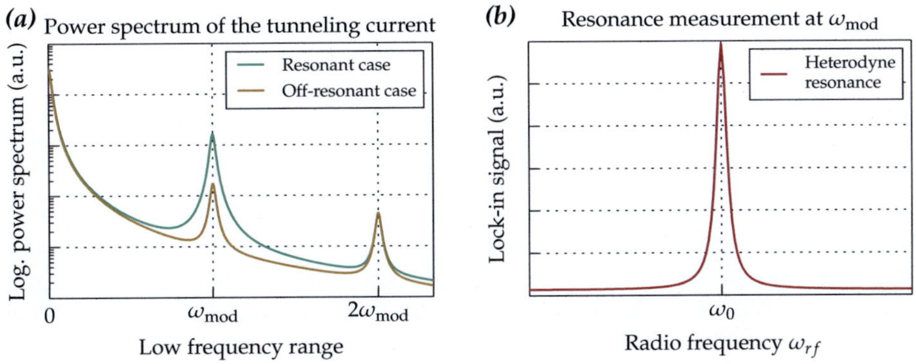

Figure 3.8: *Plot of the heterodyne signal detected by a Lock-in as described in the text. An ideal measurement with a constant phase relation (φ), a constant rf-amplitude (U_{rf}) and finite values for U_0, Λ and $\partial I(U_0)/\partial U$ were assumed. Subfigure* ***(a)*** *shows the lower part of the Fourier power spectrum of the tunneling current expansion up to second order. The sharp modulation frequencies of the Lock-in were replaced by Lorentzian lineshapes for illustration. The component at ω_{mod} shows a clear difference between the resonant and off-resonant case. The component at $2\omega_{\mathrm{mod}}$ does not depend on the resonance. In subfigure* ***(b)*** *a intended resonance curve obtained by the Lock-in is shown. The radio frequency ω_{rf} is swept and the current component at ω_{mod} is recorded. A resonance occurs at ω_0. A finite broadening due to damping was assumed.*

The second term adds a static contribution, which will be present also in the off-resonance case. As the term is proportional to the second derivative of the IV-curve it will only be present in the case of a nonlinear IV-curve. Furthermore it is proportional to U_{rf}^2, the amplitude of the radio frequency voltage. At first glance this parasitic term disrupts the resonance measurement as it can produce signals at all frequencies. But as will be shown later, it can actually help to improve the measurement. It can be used to determine the actual amplitude of the radio frequency at the tunneling junction, as the latter is perturbed by transmission effects. The determination of the transmission over frequency will be shown in chapter 4.3.2.

Figure 3.8 illustrates the previous expansion. Part 3.8*(a)* shows the lower range (dimension of the modulation frequency ω_{mod}) of the current power spectrum. These are the frequencies that will be detectable through the IV-converter. The expansion of the current up to second order was used (equ. 3.11 and 3.12). The sharp frequency lines of the Lock-in modulation were replaced by Lorentzian line-shapes for illustration. The resonant (green curve) and the off-resonant case (brown curve) are shown. As can be seen, the resonant case produces a higher signal at the Lock-in modulation frequency ω_{mod}. But there is still a peak in the off-resonant case due to the static term in equation 3.13. The frequency component at $2\omega_{\mathrm{mod}}$ will never be proportional to the resonance term; hence, the peak does not change between resonant and off-resonant case.

It is noteworthy that for expansions to higher orders (only second order is shown) all even terms will be static and all odd terms will contain a resonance component.

Subfigure 3.8***(b)*** shows a frequency sweep, which is the design template for the experiments. Either the radio frequency ω_{rf} or the frequency of the oscillating magnetic moment ω_0 has to be traversed. In case they are identical, a resonance peak occurs in the Lock-in signal (red curve). It should be noted that the shown ideal curve presumes a constant phase difference ($\varphi = 0$) between voltage and conductance. Additionally a finite linewidth of the resonance was assumed.

The two parameters phase difference (φ) and amplitude of the oscillation (A) will strongly influence the signal strength and shape. They will rely on the sample and on the excitation of the latter. In the following section, the excitation mechanisms will be discussed. This allows statements on these two parameters in the chapters corresponding to the individual samples.

3.2.2 Excitation of magnetic resonances in STM

At the beginning of this chapter it was mentioned that a strong signal over the whole measurement time is needed. The measurement time, in the case of a Lock-in assisted measurement with modulation frequencies of around $5\,\mathrm{kHz}$, is on the order of $100\,\mu\mathrm{s}$. This timescale is well above the relaxation times of most nanometer-scale magnetic moments. Hence, the oscillation might already be decayed during the measurement and a weak or no signal will be observed. Therefore, a continuous excitation of the magnetic moment is required.

Field driven excitation

In usual magnetic resonance experiments a continuous excitation is achieved through the irradiation of electromagnetic waves. The samples' magnetic moments are coupled to the oscillating magnetic field of these waves (ch. 2.1.1 and 2.1.2). This is called field coupling.

This coupling is also present in the designed STM experiment: As shown in the previous chapter, the detection of the oscillating conductance (eq. 3.2) can be solved by the introduction of a radio frequency bias voltage (eq. 3.3). The radio frequency part of the voltage will produce an oscillating current across the tunneling junction. This ac-current is a composition of a tunneling part and a displacement current. Both parts will create an Oersted field, which will curl around the tip according to Maxwell-Ampère's law:

$$\oint_{\mathrm{C}} \mathbf{H}_{\mathrm{exc}} \cdot dl = I_{\mathrm{rf}} \qquad \Rightarrow \qquad |\mathbf{H}_{\mathrm{exc}}| = \frac{I_{\mathrm{rf}}}{2\pi r} \tag{3.14}$$

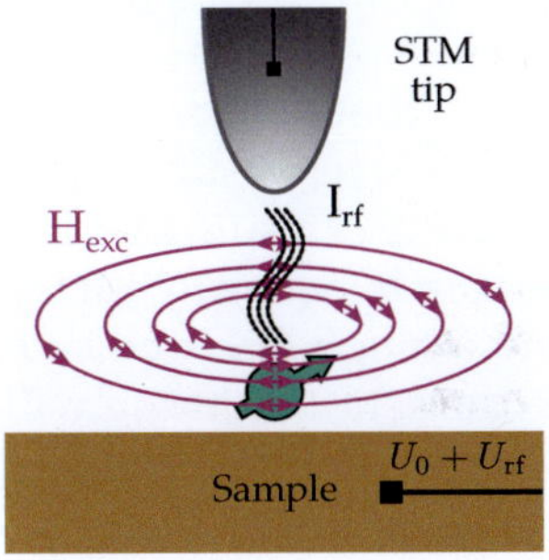

Figure 3.9: *Sketch of field excitation in STM. The radio frequency bias voltage (U_{rf}) causes a radio frequency current (I_{rf}) between tip and sample. This rf-current produces a magnetic Oersted field (H_{exc}), which is curling around the tip (purple). This field can excite the magnetic moment.*

Here, $\oint_{\mathrm{C}}$ is a closed line integral around the surface C, with dl being an infinitesimal part of the curve. Figure 3.9 shows the sketch of field excitation in the STM geometry. The generated field will excite the magnetic moment according to equation 2.6, respectively 2.19.

Since the tip radius is typically on the order of 1 µm [75], a plate capacitor for the displacement current can be assumed. Even if the displacement current is on the order of 1 mA (1 mV, 1 GHz and 1 nF capacitance of the tunneling junction), its magnetic field, at an atomic distance to the apex, is still 5 orders of magnitude lower than the field produced by the tunneling current. Therefore, only the field produced by the tunneling current will be considered further on. The field produced by a tunneling current of 100 nA at an atomic distance of 1 Å is approximately 160 A/m, which equals 0.2 mT in vacuum. This is a rather small field, but to produce such a field through an ideal coil around the tip apex with a radius of 1 mm, already a current of 1 A would be necessary. An external excitation can therefore be discarded in comparison to the field produced by the tunneling current.

It should be mentioned that field coupling can only occur when a broken rotation symmetry of tip and sample is given. This can be understood, as the field is curling around the tip apex and has to be zero at the symmetry point.

Excitation by spin transfer torque

Due to the radio frequency voltage, there is a second mechanism which will excite magnetic moments inside the tunneling junction. This is the spin transfer torque effect. It will be shown later that it predominates field excitation in the STM geometry.

The spin torque effect was independently proposed by Slonczewski [76] and Berger [77] in 1996. It can be understood when considering a spin-polarized current flowing through a ferromagnetic layer. Figure 3.10***(a)*** shows a sketch of two ferromagnetic layers to illustrate the effect. In case electrons traverse this structure, they scatter inelastically along with transfer of angular momentum. As the left layer is supposed to be fixed, momentum transfer is neglected with the outcome of a spin filtering effect. Thus the fixed layer polarizes the electrons traversing in between the layers. If the spin relaxation time of these electrons is long enough, the free layer will be exposed to this spin-polarized current. Here, inelastic scattering will produce a net torque on the magnetic layer. For a charge current flowing from the fixed to the free layer a parallel alignment is favored, whereas a charge current from the free to the fixed magnetic layer evokes an antiparallel alignment. This was experimentally verified by Myers *et al.* in 1999 [78]. It is noteworthy that although this effect is quite new it already made its way into electronic devices [79].

The effect of a spin-polarized current found by Slonczewski and Berger was generalized by Zhang *et al.* in 2002 [80,81]. They showed that a spin-polarized current not only produces the spin torque term but also an effective field term (also called field-like torque). Regarding a ferromagnetic situation, these torques can be described as an extension to the Landau-Lifshits-Gilbert equation (2.18):

$$\frac{d\mathbf{M}}{dt} = \ldots + \underbrace{a\,(I)\ \mathbf{M} \times \left(\hat{\mathbf{x}}' \times \mathbf{M}\right)}_{\text{spin torque}} + \underbrace{b\,(I)\ \mathbf{M} \times \hat{\mathbf{x}}'}_{\text{effective field}}\,. \tag{3.15}$$

Here, a and b are the driving torques and $\hat{\mathbf{x}}'$ is the unit vector giving the direction of the spin-polarized current. Figure 3.10***(b)*** shows the effect in the context of the LLG equation describing a precessing magnetic moment. As can be seen, the effective field term is perpendicular to the magnetization, pointing in the direction of the precession. The spin torque term in contrast can counteract the damping. Of course, the directions of the torques depend on the spin polarization $\hat{\mathbf{x}}'$ and on the sign of the parameters a and b. As the signs of a and b depend on the direction of the current, it is possible to generate a spin torque which is continuously compensating the damping, and thereby exciting the system. This can be achieved by means of an ac-current with the frequency ω_{rf} adapted to the precession frequency ω_0.

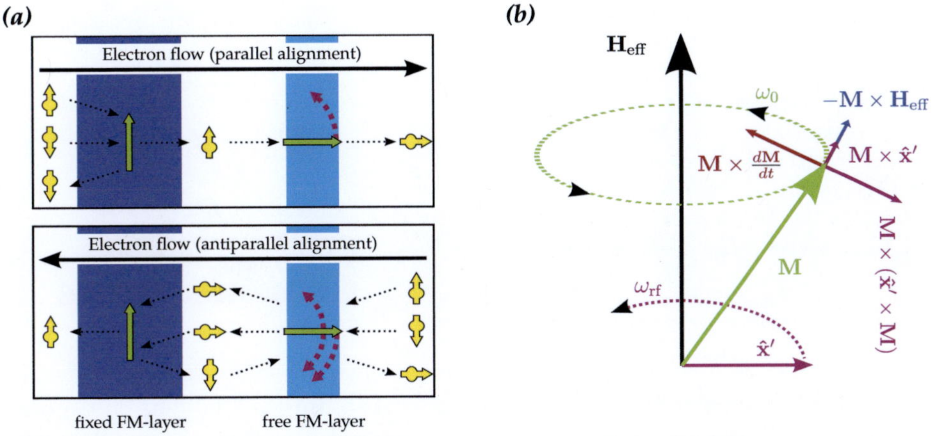

Figure 3.10: ***(a)*** *Spin transfer torque in two ferromagnetic layers (blue), with their spin polarization at the fermi energy (green arrows). The electrons (yellow) carry a spin, whose probability to be found in the pure up/down state is represented by the direction of the yellow arrows. The torque acting on the free layer is depicted in purple. For a charge current from the fixed- (free-) to the free- (fixed-) layer a net torque towards a parallel (antiparallel) alignment is acting on the free layer.* ***(b)*** *Spin transfer torque and effective field acting on a precessing magnetic moment* $\mathbf{M}$. $\hat{\mathbf{x}}'$ *is the direction of the spin-polarized current. It produces the spin torque term* $\mathbf{M} \times (\hat{\mathbf{x}}' \times \mathbf{M})$ *and the effective field* $\mathbf{M} \times \hat{\mathbf{x}}'$. *A continuous excitation can be achieved by an oscillating* (ω_{rf}) *spin-polarized current.*

Figure 3.11: Spin transfer torque in the STM geometry. The spin-polarized tip (light green) acts as a spin filter producing a spin-polarized current. This current will yield a torque (purple) on the magnetic moment. Depending on the sign of the current, a parallel or an antiparallel alignment is pursued. A radio frequency bias voltage (U_{rf}), which will cause a radio frequency current (I_{rf}), can resonantly excite the system.

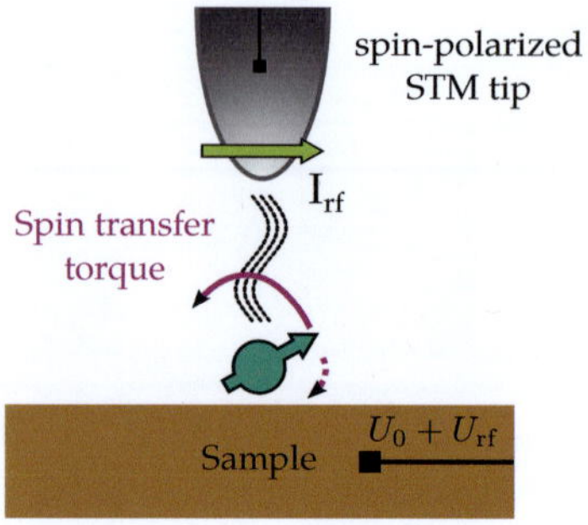

In the STM geometry the situation corresponds to figure 3.11. The spin-polarized tip can be assumed as a fixed layer which will act as a spin filter. The torque affecting the magnetic moment on the sample will therefore either favor a parallel or an antiparallel alignment. This depends on the sign of the radio frequency current (I_{rf}). This is exactly the kind of excitation that is needed to maintain an oscillation of the magnetic moment. The strength of the excitation depends on the spin polarizations of tip and sample and on the magnitude of the rf part of the tunneling current induced by the applied rf-voltage.

Comparison: Field excitation - Spin transfer torque

To get an idea of which of these two excitation mechanisms is dominant in the STM geometry, it is natural to compare their strengths by a simple estimation. This can be done by comparing the corresponding forces acting on the magnetic moment. For field coupling this is the Oersted field produced by the tunneling current. And for the spin-torque term it is the torque term given by Slonczewski [76]. The ratio between both is given by:

$$\frac{\tau_{\mathrm{torque}}}{\tau_{\mathrm{field}}} = \frac{I_{\mathrm{rf}} e^{-1} g\,(\theta, P) \sin\theta}{I_{\mathrm{rf}} \gamma \mu_0 \,(2\pi r)^{-1}} \tag{3.16}$$

Here, g is a function scaling of the spin torque (see [76]). It depends on the magnitude of the spin polarizations $P_{\mathrm{T}} \cdot P_{\mathrm{S}}$ and on the angle θ between them. e is the electron charge and μ_0 the vacuum permeability. I_{rf} is the tunneling current and γ the gyromagnetic ratio. This ratio depends on the sample. The free electron has a value of $\gamma_{\mathrm{e}} = 28\,{}^{\mathrm{GHz}}\!/_{\mathrm{T}}$, whereas for a ferromagnetic sample values of about $\gamma_{\mathrm{fm}} = 0.66\,{}^{\mathrm{GHz}}\!/_{\mathrm{T}}$ are realistic (see chapter 5). The radius r can be approximated by a mono-atomic tip $r = 1\,\text{Å}$. A typical value for the spin polarization, e.g. iron ($P_{\text{Fe}} = 0.4$), can be used. The angle θ, entering the scaling of the spin torque, influences the ratio quite radical: At $\theta = 0$ the spin torque will vanish and the excitation has to relay on field excitation. But already at small angles of about $\theta = 18°$ between the tip polarization and the magnetic moment, the ratio is in the range of $\tau_{\text{torque}}/\tau_{\text{field}} \approx 2 \times 10^2$ for an electron and $\tau_{\text{torque}}/\tau_{\text{field}} \approx 10^4$ for a ferromagnet. It can be concluded that the excitation in the designed magnetic resonance experiments is a combination of both effects. In parallel alignment field excitation should be present. But as soon as a tiny oscillation amplitude is accomplished, the spin torque effect should dominate the excitation.

Furthermore, it is noteworthy that the phase relation φ between the voltage and the oscillation of the magnetic moment is defined by the excitation. This should profoundly influence the experiments outcome as the heterodyne detection depends on the this phase (eq. 3.13). The details of these influences will be discussed in the chapters corresponding to the individual samples.

3.2.3 Summary of the designed measurement technique

A new magnetic resonance technique inside an STM was introduced. It is based on heterodyne detection, which is achieved by a spin polarized tip and the introduction of a radio frequency modulation of the bias voltage. This allows not only to detect the oscillations of a magnetic moment but will also resonantly excite the latter via the spin torque effect.

As this is a new *experimental* technique, a proof of concept measurement is needed to show the feasibility of the designed concept. To boost the chances of a successful measurement, many requirements with regard to the sample and the experimental setup have to be met. The main novelty in the setup is the introduction of the radio frequency bias voltage. This is associated with transmission problems and will be tackled in the chapter of the experimental setup (ch. 4).

The requirements regarding a test sample are also very demanding, which impedes the search for a suited specimen. The main target is always to achieve a strong signal (eq. 3.13) compared to experimental noise. As the signal depends on the amplitude of the oscillation, the excitation must be strong enough to drive the system under study. This demands low anisotropies and a small damping value of the sample. The phase relation φ has to be considered also for the estimation of the signal strength. To be able to predict it, an easy simulateable system is desired. The simulations could not only be used to estimate the signal shape and size but also help to predict the expected resonance frequency. A sample with a resonance frequency in the experimentally accessible range has to be chosen.

The last element to mention is the spin-polarized tip. Empirically, it is very challenging and insecure to prepare/achieve a strongly spin-polarized STM tip. But the spin polarization of the tip directly scales the signal of the resonance measurement. Thus, a sample which allows *in situ* confirmation of the spin polarization would improve the probability for a successful measurement a lot. Furthermore, the assumption that the spin polarization of the tip is fixed with respect to the sample (ch. 3.2.2) has to be taken into considerations. This is strongly fulfilled for antiferromagnetic tips, where the exchange interaction is very strong. But in ferromagnetic tips this is hard to predict as the exact tip geometry is unknown. As ferromagnetic tips additionally exhibit a stray field, which will influence the resonance, antiferromagnetic tips are strongly preferable [82]. But the experience shows that good, stable antiferromagnetic tips are very hard to achieve.

If all these requirements can be fulfilled, it is highly probable to successfully prove the new measurements concept and its high prospects. The test samples themselves and their advantages and disadvantages will be approached in the corresponding chapters 5, 6 and 7.

3.2.4 Equivalent solid state experiment - The spin torque diode

In 2005, Tulapurkar *et al.* showed an experiment which is closely related to the heterodyne method of the present work [83]. They called the observed behavior spin torque diode effect. The key difference is the use of a solid state device instead of an STM. They used an array of artificially created nano-pillars which consist of a magnetic tunneling junction. A simplified sketch of one nano pillar is shown in figure 3.12.

The nano pillars consist of a free (light blue) and fixed (dark blue) ferromagnetic layer. They are separated by an insulating, crystalline MgO barrier (red), which yields a huge TMR. This structure is equivalent to the situation of spin-polarized STM. Tulapurkar *et al.* now applied a radio frequency current through the tunneling junction. This current will excite a precession of the free magnetic layer due to spin transfer torque. Hence, the way of excitation equals the approach of this work, which was explained earlier (ch. 3.2.2). If the frequency of the current equals the precession frequency of the free ferromagnetic layer, the heterodyne mixing of current and oscillating TMR will yield a constant voltage U_{DC}. This voltage was measured for frequencies between 3 and 11 GHz. With this, also the detection is similar to the concept in STM (ch. 3.2.1). The magnetic origin of the signal was confirmed by applying an external magnetic field. This field changes the resonance frequency of the free layer, which was observed experimentally.

The experiment and technique by Tulapurkar *et al.* is consequently the equivalent of the experiments presented in this work using a solid state tunneling junction. The outstanding advantage by using an STM is, of course, its lateral resolution. This allows to measure the effect not only on an array of nano objects but on a single one.

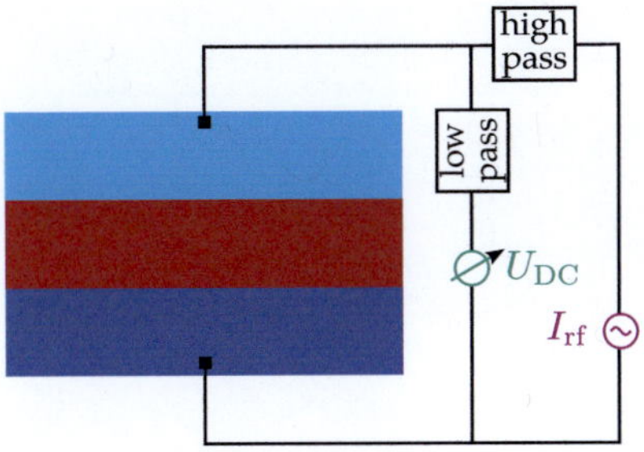

Figure 3.12: *Simplified sketch of a solid state experiment after [83]. A magnetic tunneling junction consisting of a free (light blue) and fixed (dark blue) ferromagnetic layer is shown. They are separated by an insulating MgO barrier (red). A radio frequency current (I_{rf}) is send through this structure. In case the frequency of the current equals the induced oscillations of the free magnetic layer, a constant voltage U_{DC} is generated because of heterodyne mixing.*

4 Experimental setup

The experiments shown in this work were carried out in a scanning tunneling microscope. As this study is based on a new measurement concept, the extension of an exiting, "conventional" low temperature STM setup is part of this work. The features and technical details of the conventional setup are explained first. Afterwards, the basics of transmission lines for high frequency electromagnetic waves are presented. They are needed to guide the radio frequency voltage to the tunneling junction, which is an essential component of the new concept. Subsequently, the modifications of the setup are outlined. Finally, transmission measurements are presented, which confirm the modification of the experimental setup for radio frequency voltages and show the performance of the changes.

4.1 Conventional low temperature STM setup

Simple STM setups are used in almost all laboratories or even student lab courses, which target surface science nowadays; like the first STMs they work at room temperature and often even at ambient pressure. It will be explained in the two following sections that these two preconditions significantly limit the prospects of an STM.

Thus, the setup used in this work is an STM in ultra high vacuum (UHV) held at 4.2 K, the boiling temperature of liquid helium. These two constituent properties entail many technical issues that need detailed solutions. Figure 4.1 shows the whole setup which is essentially confined by the vacuum chambers. It consists of an STM chamber (green), containing the cryostat operated at 4.2 K. The preparation chamber (blue) is equipped with an Ar sputter gun, evaporators for metal thin films, and molecule deposition and fundamental sample characterization tools. This allows *in situ* preparation and analysis of the samples and tips for STM experiments. Both, samples and tips, can be introduced into the vacuum chambers through a two stage load lock (red).

In the following sections the different concepts and parts of the setup will be established and explained. Until the start of this work, the setup itself was mainly dedicated to spin-polarized STM measurements [85–87]. It was therefore optimized for low noise in direct current operation. During the present work, it has been modified for radio frequency measurement as will be shown later (ch. 4.2.2).

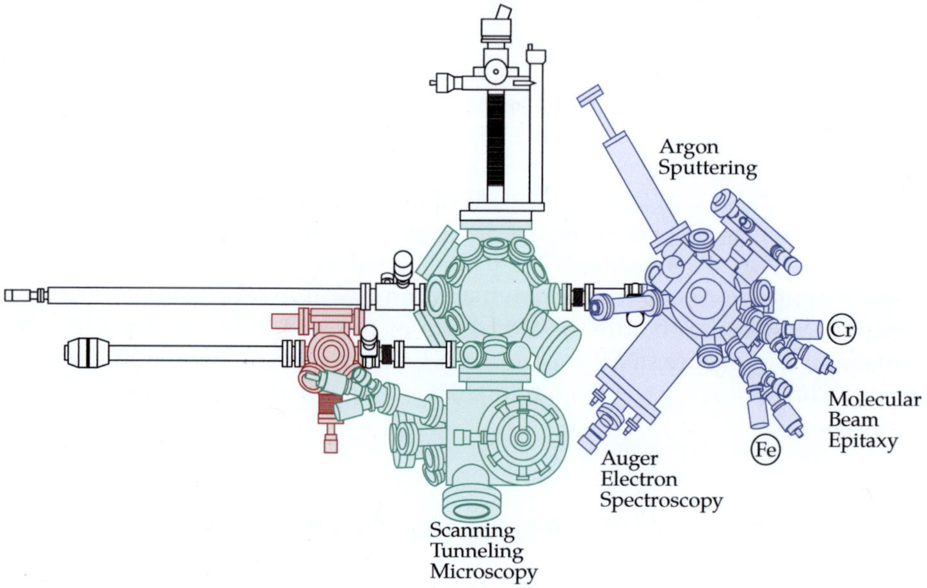

Figure 4.1: *Top view of the complete STM setup (modified according to [84]). The major areas are illustrated in different colors: The main chamber with cryostat and housing of the STM is depicted in green. It is connected to the blue part, the preparation chamber. The latter contains all necessary parts for the cleaning of the samples and the growth of nanostructures. There are also fundamental analysis methods available. Samples or STM tips can be inserted through the load lock, which is depicted in red.*

4.1.1 Vacuum

For examining surfaces and nanostructures on the atomic scale, it is import to work in an environment clean of contaminants. In an experiment on the atomic length scale even a single foreign atom can distort or destroy the measurement. Such foreign atoms can be adsorbed from the surrounding of the sample. To avoid such adsorption, the sample is held, prepared and measured in a vacuum.

Vacuum is defined as the absence of matter in any thermodynamic phase. The atmosphere of the earth consists of nitrogen, oxygen and many other components at a pressure of about 1024 mbar. Since the particles of the air can move around, they will hit the sample surface and can be adsorbed as mentioned before. The amount of particles at atmospheric pressure is so high that a sample surface of 1 mm^2 is fully covered with adsorbates in less than 1 ns (for common adsorption coefficients). This is far too fast to allow reasonable measurements. Therefore, the entire preparation and the experiments are carried out in a chamber which is air-evacuated as good as possible. In the present setup a pressure of about 10^{-10} mbar was achieved. This range is defined as ultra high vacuum (UHV). It is sufficient to keep samples clean for measurements over days. In the following paragraph the technical utilities needed to achieve UHV are explained.

In order to reach these low pressures, several different types of vacuum pumps are used in series or in parallel. The first stage is a rotary pump, which reduces the pressure down to about 10^{-2} mbar. This is sufficient for the next stage, the turbo molecular pumps (TMP). They need a low base pressure because they solely work in the molecular regime, i.e. the mean free path of the gas molecules has to be on the order of 10 mm. Their working principle is based on momentum transfer to the molecules by a rotor and a stator. Many rotor/stator pairs in series allow pressures of about 10^{-9} mbar.

Rotary and turbo molecular pumps cause mechanical vibrations. This is disruptive for STM operation; hence, they must be turned off during measurement. To maintain the vacuum during this time, two types of chemical pumps are used. The first are titanium sublimation pumps, where a high current is send through a titanium wire. The sublimed titanium condenses on the chamber walls and chemically binds residual gas molecules. The second type are ion-getter pumps. These ionize gas molecules and a strong electric field accelerates them towards a reactive surface. There the molecules get buried deep inside a reactive metal. It is obvious that these pumps do not work for noble gases since they do not react with other materials.

At last, the low temperature cryostat, which will be explained in the following section, must be mentioned. Its metal shields act as a pump, too, since most residual gases are frozen out at the cryostat's operation temperature (4.2 K).

As desorption of gas molecules from the vacuum chamber walls is significant at pressures lower than 10^{-6} mbar, simple pumping of the volume is not sufficient. Therefore, the chamber is made of stainless steel and can be heated to several hundred degrees Celsius. This "bake-out" process increases the desorption from the chamber walls, and thus speeds up the air-evacuation to about two days. Thereafter, UHV of about 10^{-10} mbar can be reached. The measurement of such low pressures is done using ionization gauges.

4.1.2 Low temperatures

For many STM studies low temperatures are essential or at least helpful. At room temperature the thermal energy is about $E_{\mathrm{therm}} \approx 25\,\mathrm{meV}$. This energy is already higher than most diffusion barriers on metallic surfaces so that no stable configuration is reached. This means that molecules or atoms can move on the sample surface and the measurements get distorted. In principle any kind of thermal drift has to be prevented to allow highly sensitive measurements. This is especially true for spectroscopic investigations where long integration times are needed to achieve a satisfactory signal-to-noise ratio. Another reason for working at low temperatures is thermal energy broadening. At finite temperatures the tunneling electrons in an STM experiment follow the Fermi-Dirac distribution (ch. 2.2.1). This means that their energy is only confined up to a certain value, which might be too high to resolve low energy features, e.g. in a spectroscopy experiment. The last reason for STM operation at low temperatures are the magnetic properties of the sample. As the investigation of ferromagnetic samples is intended, the operational temperature has to be below the sample's Curie temperature. For example the magnetic skyrmion structure (ch. 6) only exists at low temperatures ($T < 27.8\,\mathrm{K}$ [88]).

For the reasons mentioned above, the STM is mounted at the bottom of a liquid helium cryostat. The latter consist of a $4\,\mathrm{L}$ tank surrounded by shieldings and a liquid nitrogen tank. The shieldings and the nitrogen tank reduce the liquid helium consumption to about $50\,\mathrm{ml/h}$. This ensues a standing time of about three days. To access the STM for sample or tip transfers, the shieldings have doors that can be opened and closed by a manipulator.

4.1.3 Sample preparation and characterization

Sample preparation and characterization takes place in a separate chamber. This is necessary since the preparation process requires the inlet of particular gases, which would pollute the STM chamber. Therefore, the preparation chamber is equipped with TMPs of their own; so that it can be brought back to UHV conditions after a preparation process.

Preparation

The first step of a usual preparation process is ion-etching of the sample, so-called sputtering. A noble gas, typically argon, is led into the chamber, ionized and accelerated onto the surface by a sputter gun. The energy, of usually about $3\,\mathrm{keV}$, is sufficient to remove entire impurity clusters from the sample. Through this treatment the sample surface gets very rough since the argon ions carry so much momentum that not only impurities but also atoms of the sample surface are shot away. To cure the sample, it is heated to temperatures well below the melting point. The elevated temperatures provide enough energy to the surface atoms to allow them to overcome diffusion energy barriers and move to their ground state. The heating of samples up to 2200 °C is realized by electron bombardment of the specimen. The temperature is measured by a commercial pyrometer.

A second type of cleaning process, which is available in the preparation chamber and used for this work, is glowing in oxygen atmosphere. 99.995 % pure oxygen is introduced into the chamber through a leak-valve to a pressure of about 5×10^{-7} mbar. Once, e.g. tungsten is heated in this environment, carbon, the main impurity in tungsten, segregates to the surface and reacts with oxygen to form CO and CO_2. These gases desorb from the surface; hence, a net cleaning is achieved.

To fabricate the necessary nanostructures, the chamber is equipped with metal and molecule evaporators. These evaporators are based on molecular beam epitaxy (MBE); thus, low growth rates at high purity are attained. The metal evaporators consist of a pure metal rod or a molybdenum crucible filled with metal flakes. They are heated by electron bombardment until a sufficient vapor pressure is reached. The detached atoms are deposited onto the target. The surroundings of the evaporator are water cooled to prevent contamination with foreign materials. Since the evaporation rate has to be controlled, a flux-meter is mounted at the outlet of the evaporator. The organic molecule evaporator is a slightly simpler construction. An alumina crucible is heated by a filament while the temperature is monitored by means of a thermocouple. The thermally vaporized molecules are deposited on the sample. The evaporation rate is solely controlled by the heating power.

The preparation methods explained are very general. The actual processes needed to prepare the individual samples are explained in the chapters corresponding to the samples. A good review of preparation methods of different materials was published by Musket *et al.* [89].

The preparation of STM tips is done by applying the same methods. In this work only tungsten tips were used. A polycrystalline tungsten wire was electrochemically etched and then inserted into the UHV chamber. Afterwards, the aforementioned sputtering and heating treatment was applied. This yields sharp and usually very stable tips for high quality STM measurements. For spin-polarized measurements the tungsten tips were coated with anti-/ferromagnetic materials. This can be done in the preparation chamber using the metal evaporators. Fe, Cr and Mn were used during this work. An annealing treatment after deposition is often helpful to ensure proper growth of the magnetic material on the tip. For Fe coated tips this is not only helpful but essential to achieve spin polarization (heating the tip to approximately 500 °C for 2 s is sufficient). The direction of the spin polarization can be controlled by usage of different magnetic materials and different coverages. For Fe an in-plane spin polarization is highly probable, but sometimes canted directions were obtained. When using antiferromagnetic Cr the direction is thickness-dependent (25 to 45 ML yield out-of-plane polarization; higher coverage gives in-plane polarization) [75]. For Mn the spin polarization is unknown and most likely random.

Figure 4.2: *Homebuilt STM. Made of gold coated copper. The sample is inserted in the stage at the bottom, which is equipped with a coarse motion in x-direction. The scanner tube piezo is located in the middle, inside a copper prism. The whole prism is part of the coarse motion in z-direction. Inside the cryostat the STM is attached to springs (not visible) for vibration damping.*

Sample characterization

Since the transfer to the STM, the cooling down and the approach into tunnel contact are rather time consuming, the preparation chamber is equipped with some sample characterization experiments for immediate verification of the preparation process. Auger electron spectroscopy (AES) is available, which is a tool for chemical analysis of the sample surface. It measures the amount of secondary auger electrons over energy. These spectra can be compared with reference data to specify the chemical composition. There is the possibility to measure low energy electron diffraction (LEED), as well. This allows the determination of the crystallographic structure of a substrate and of the deposited nanostructures.

4.1.4 The STM head

The main component of the whole setup is the homebuilt scanning tunneling microscope (fig. 4.2). For being non-magnetic, mechanical stability and thermal conductivity the STM body is made of gold coated copper. Since elimination of mechanical vibrations is crucial in scanning tunneling microscopy, the whole STM body is attached to CuBe springs. They are designed to have the lowest resonance frequency possible. Additionally, the complete vacuum chamber is placed upon air dampers, which will suppress vibrations caused by people walking around the machine etc. The STM consists of an upper part, the housing of the scanner tube, and a lower part, the sample stage. The scanner tube is fixed to a copper prism, which can be moved by piezo slip-stick motors for coarse motion in z-direction. The sample stage is a rack, also made of copper, located exactly below the scanner tube. It is equipped with ruby balls on top of piezo actuators for coarse motion in x-direction. For normal operation, the sample is inserted into the rack and the tip is approached by means of the z-coarse motion. If an inadequate area of the sample is caught, the scan area can be changed on the macroscopic scale by means of the x-coarse motion without external manipulation. While being in tunneling contact, the scan motion is solely conducted by the tube piezo.

Since the tip must be prepared and replaced in situ, the STM possesses a tip exchange mechanism. A special sample plate which can hold the tip by means of a spring is inserted into the STM. The z-coarse motion is lowered to grab the tip. Then the x-coarse motion releases the tip from the spring and the z-coarse motion is free to retract the exchanged tip.

4.1.5 STM electronics

The STM electronics used in this work is a commercial NANONIS™ system. It is connected to an ordinary computer for scan commands and data acquisition. The electronics determines and generates all required voltages and applies them to the piezo elements. To measure a tunneling current in the pA range, a FEMTO® DLPCA-200 transimpedance amplifier is used. The provided gains of 10^7, 10^8 and 10^9 V/A were used. Since for stability, the finite input capacitance and the back coupling of the operational amplifier have to be compensated by small internal capacitors, the amplifier's bandwidth is limited to 50, 7 and 1 kHz, respectively. The resulting signal is read by the NANONIS™ electronics for data acquisition. For constant current mode, the electronics features a z-feedback mechanism. The tunneling current is used as the input of the feedback which adjusts the distance between tip and sample (e.g. the voltage of the z-piezo) to keep the tunneling current constant.

For spectroscopic measurements an AMATEK 7320 DSP digital Lock-in amplifier is used. The preamplified tunneling current is supplied to its input allowing to measure the modulated component of the tunneling current at a reference frequency ω_{mod}. Depending on the quantity that is modulated, different experiments like STS (modulation of the bias voltage, ch. 2.2.2) or magnetic resonance measurements (amplitude modulation of the radio frequency voltage, ch. 3.2.1) can be performed. The signal measured by the Lock-in is fed to an auxiliary input of the NANONIS™ electronics.

It should be mentioned that the entire "conventional" STM electronics were optimized for low noise and low frequencies. This means that cables and connections used for the circuits show good performance in noise at low frequencies, but they are not suited for frequencies above 200 kHz. With increasing frequencies of voltages and currents, the wavelengths of these electromagnetic waves become smaller and the geometrical layout of the circuits becomes relevant. The necessary optimizations are described in the following chapter.

4.2 Modification of the STM setup for small timescales

To achieve the experimental requirements for measurements according to the principle explained in chapter 3, the whole electronic composition of the STM has to be considered in a different view. Normal STM operation is based on voltages and currents at a time scale between DC and 50 kHz. The frequencies of magnetic resonance techniques, however, lie in the range of several 100 MHz up to 10 GHz. Voltages and currents in this high frequency range are called ultra high radio frequencies (UHF) or microwaves and cannot be treated like a standard circuit problem. Therefore, the concept of transmission lines and the generation of radio frequencies will be explained in the following sections. As mentioned before, all cables in the STM setup were optimized for low noise, and thereby for low frequencies. A modification of the STM's cables and connections is therefore necessary. But not all parts of the STM and the microwave world are compatible so that a compromise has to be found.

4.2.1 Transmission line theory

Electromagnetic waves, in general, are described by Maxwell's equations. At low frequencies the equations simplify to standard circuit theory. But with rising frequencies the wavelength shrinks down in size. If the wavelength is on the order of the circuit size, the phase of voltage and current changes significantly over the circuit or the components physical extent. In this range transmission line theory applies. At even higher frequencies the limes of geometrical optics reduces complexity.

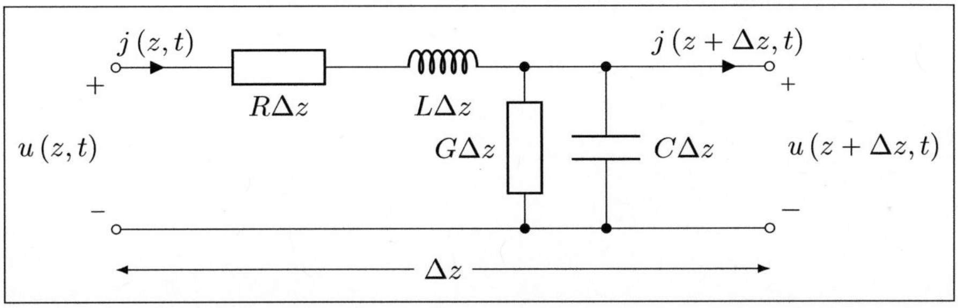

Figure 4.3: Lumped-element representation of an infinitesimal part Δz of a microwave transmission line. R (resistance in series), L (inductance in series), G (shunt conductance) and C (shunt capacitance) are per-unit-length quantities. The characteristics of voltage $u\left(z,t\right)$ and current $j\left(z,t\right)$ over the transmission line are the quantities of interest.

A simple infinitesimal schematic representation of a transmission line is shown in Figure 4.3. In reality this could, for example, be a coaxial cable. R is the resistance of the wires, L the self inductance, C the shunt capacitance and G the shunt conductance. Kirchhoff's laws for voltage $u\left(z,t\right)$ and current $j\left(z,t\right)$ yield:

$$u\left(z,t\right)-R\Delta zj\left(z,t\right)-L\Delta z\frac{\partial j\left(z,t\right)}{\partial t}-u\left(z+\Delta z,t\right)=0, \qquad (4.1a)$$

$$j\left(z,t\right)-G\Delta zu\left(z+\Delta z,t\right)-C\Delta z\frac{\partial u\left(z+\Delta z,t\right)}{\partial t}-j\left(z+\Delta z,t\right)=0. \qquad (4.1b)$$

By dividing through Δz and taking the limes as $\Delta z\rightarrow 0$, it is easy to find the *telegrapher's equations:*

$$\frac{\partial u\left(z,t\right)}{\partial z}=-Rj\left(z,t\right)-L\frac{\partial j\left(z,t\right)}{\partial t}, \qquad (4.2a)$$

$$\frac{\partial j\left(z,t\right)}{\partial z}=-Gu\left(z,t\right)-C\frac{\partial u\left(z,t\right)}{\partial t}. \qquad (4.2b)$$

It is common to separate the time dependence first ($u(z,t) = U(z)\,e^{i\omega t}$ and $j(z,t) = I(z)\,e^{i\omega t}$). Stationary solutions of wave propagation on the transmission line can be found for the remaining differential equations. A reasonable ansatz is:

$$U(z) = U_0^+ e^{-\gamma z} + U_0^- e^{\gamma z}, \tag{4.3a}$$

$$I(z) = I_0^+ e^{-\gamma z} + I_0^- e^{\gamma z}. \tag{4.3b}$$

This yields a complex propagator constant:

$$\gamma = \alpha + i\beta = \sqrt{(R + i\omega L)(G + i\omega C)}. \tag{4.4}$$

The coefficients U^+ and I^+ (U^- and I^-) represent the part of the waves propagating in $+z$ ($-z$) direction. By applying the obverse stationary telegrapher's equation (eq. 4.2b) to the voltage solution (eq. 4.3a) and comparison to the current solution (eq. 4.3b) a relation analogous to Ohm's law is found:

$$I(z) = \underbrace{\frac{\gamma}{R + i\omega L}}_{=:\frac{1}{Z_0}} \left(U_0^+ e^{-\gamma z} - U_0^- e^{\gamma z}\right), \tag{4.5}$$

$$\Rightarrow I(z) = \frac{U_0^+}{Z_0} e^{-\gamma z} - \frac{U_0^-}{Z_0} e^{\gamma z}. \tag{4.6}$$

The ratio of voltage to current Z_0 is the characteristic impedance of the transmission line. This intrinsic quantity strongly determines the influence of each component inside a radio frequency circuit. It is important to note that the characteristic impedance is frequency-dependent. This indicates that the transmission of electromagnetic waves through a cable is *a priori* not constant over frequency. To illustrate the influence of the impedance, an arbitrary load impedance Z_L at the end of the transmission line ($z = 0$) should be considered (fig. 4.4). Let us assume an incident wave from a generator at $z < 0$ in the form of $V_0^+ e^{-i\beta z}$ (the transmission line is assumed to be lossless). The voltage and the current at $z = 0$ are then give by:

$$\frac{U(0)}{I(0)} = \frac{U_0^+ + U_0^-}{U_0^+ - U_0^-} Z_0 = Z_L. \tag{4.7}$$

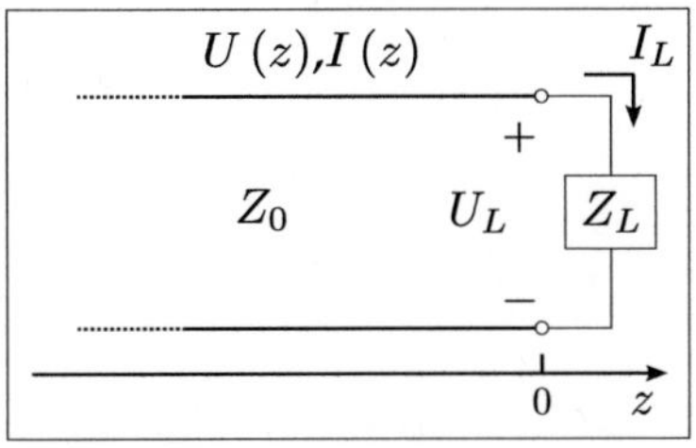

Figure 4.4: *Transmission line terminated with an arbitrary load impedance Z_L. The transmission line possesses an intrinsic impedance Z_0. An electromagnetic wave ($U(z)$, $I(z)$) generated at $z < 0$ is impinging on the circuit shown.*

When solving this equation for U_0^-, the amplitude of the reflected wave traveling in negative z direction is retrieved. Thereby, the *voltage reflection coefficient* is defined as the amplitude ratio between incoming and outgoing wave:

$$\Gamma := \frac{U_0^-}{U_0^+} = \frac{Z_L - Z_0}{Z_L + Z_0}. \tag{4.8}$$

If the load impedance equals the characteristic impedance of the line ($Z_L = Z_0$), no reflection is expected. This is called a matched impedance. In this case the magnitude of the voltage on the line is constant. If the impedance is not matched, a part of the wave is reflected and is superimposed with the incident one. This forms a standing wave. For example, an open-ended transmission line can be considered. This means $Z_L = \infty$ (finite voltage, zero current) and as a result $\Gamma = 1$. In this case the whole wave is reflected. The magnitude of the voltage along the line now depends on a phase factor $e^{i2\beta z}$. A widely used quantity is the standing wave ratio (SWR). It is defined as the ratio between the maximum and the minimum voltage on the line:

$$\text{SWR} = \frac{U_{\text{max}}}{U_{\text{min}}} = \frac{1 + |\Gamma|}{1 - |\Gamma|}. \tag{4.9}$$

For a perfectly matched line SWR $= 1$, whereas an open-ended line yields SWR $= \infty$.

As seen, a perfect match of the impedances inside a circuit is crucial for a flat radio frequency transmission. This means all components in a circuit, from the generator to the load, have to posses the same impedance. Thus, obvious mismatches, e.g. differently specified cables but also tiny differences in simple connectors between two coaxial cables, should be avoided. An open ending, like the tunneling gap of the STM, is of course the absolute worst case.

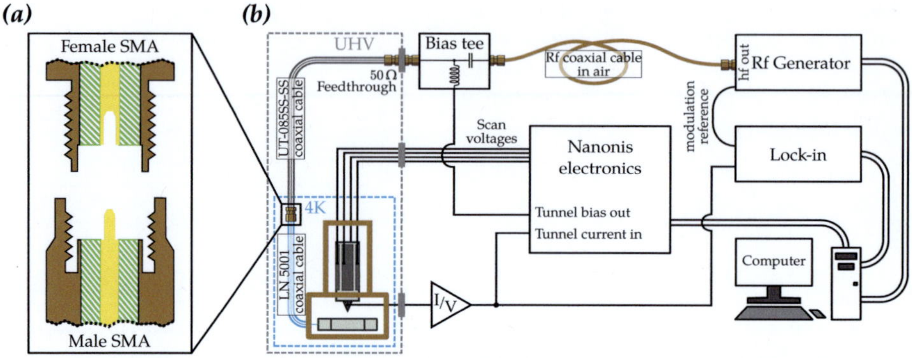

Figure 4.5: ***(a)*** *Male and female SMA connectors. They consist of stainless steel bodies (brown), a PTFE dielectricum (green) and gold coated conductors (yellow). The connectors feature* 50 Ω *impedance and are designed for frequencies up to* 18 GHz*. Male and female connectors match exactly so that no gap between phase, ground or dielectricum could cause an impedance mismatch.* ***(b)*** *Fully modified electronic setup. On the left-hand side the STM with its main electronic features is sketched. Inside the UHV chamber two new radio frequency cables were used. Outside, a bias tee combines the DC bias with the radio frequency voltage. The latter is obtained from a generator. All connections of the transmission line are realized with SMA connectors featuring* 50 Ω *impedance. The remaining components belong to a conventional STM setup.*

4.2.2 Electrical modification of the STM setup

As shown in the previous section, impedance matching from the rf-generator to the sample is very important. The STM in its original configuration was not suited for this kind of experiments. Air-to-vacuum feedthroughs, connections inside and outside of the chamber and the cables themselves were not impedance matched and exhibited to high damping values and reflection coefficients at the desired radio frequencies. For this reason a new rf transmission line was incorporated into the existing STM setup. According to the designed magnetic resonance measurements (ch. 3), only one high frequency transmission line to the STM is needed. This transmission line conducts the radio frequency voltage to the tunneling junction. As the readout is based on heterodyne detection, yielding a constant current component, the current line can remain untouched.

In the field of microwave engineering the SubMiniature version A (SMA) connectors are very common (fig. 4.5*(a)*). They feature 50 Ω impedance and are designed for frequencies from DC to 18 GHz. Their dielectricum, which governs the capacitance and the shunt resistance, is Polytetrafluoroethylene (PTFE), also known as Teflon. It is important to note that the connectors are designed to fit very accurate. This prevents gaps between them where the different material in the gap, i.e. air, would cause an impedance mismatch. As these connectors are also usable under UHV conditions, they where placed throughout the new radio frequency transmission line.

Figure 4.5*(b)* shows a sketch of the electrically modified STM setup including the new transmission line. The rf-wave-generator (ch. 4.2.3) is connected to a bias tee with a 4 m in length standard 50 Ω coaxial cable with low damping characteristics (1.45 dB/m at 2 GHz). The bias tee mixes the high frequency voltage to the conventional DC bias voltage of the NANONIS™ electronics. The following air-to-vacuum feedthrough was replaced by a 50 Ω SMA one. Inside the vacuum chamber two different cables are used. The first one is a semi-rigid coaxial cable, type *UT-085SS-SS* from Micro Coax with 50 Ω impedance and a length of about 1.8 m. It exhibits low thermal conductance because its conductors are made of stainless steel. Therefore, it was used for the transition from room temperature to 4 K. The drawback is a rather high damping of around 7 to 8 dB/m at 2 GHz. At this section an SMA connection to an *LN 5001* coaxial cable from elspec GmbH is installed (30 cm long). This cable is quite flexible which is crucial to not influence the damping springs of the STM. It is specified with low noise characteristics, 50 Ω impedance and a damping of 2.08 dB/m at 2 GHz. Thus, all components from the rf-generator down to the STM exhibit an impedance of 50 Ω and are therefore impedance matched. The total damping of the cables should be around 19 dB at 2 GHz.

Since the tunneling bias (i.e. the sample stage) has to be isolated against ground, and hence is not 50 Ω impedance matched, the transmission line has to be interrupted at some point. For that reason, the last cable was decomposed as close as possible to the bias stage and directly coupled to the latter. The distance from the impedance mismatch to the tunneling junction ($\approx$ 1 cm) is well below the wavelength of the design frequencies (up to 3 GHz $\widehat{=}$ 10 cm). This should assure sufficient radio frequency transmission into the tunneling junction.

As usual in STM, the measurement shall be improved by modulation and phase sensitive detection by means of a Lock-in amplifier (ch. 3.2.1). The modulation itself is done by the rf-generator, which features two suitable methods that are depicted in the next section (ch. 4.2.3). For synchronization, the Lock-in is referenced to the rf-generator. Thus, it is able to demodulate the tunneling current and thereby retrieve the intended signal. All three main components, the rf-generator, the Lock-in and the NANONIS™ electronics are connected to a computer which controls the measurement and reads the data. The ordinary scan software of the NANONIS™ electronics was therefore modified. This will be shown in section 4.2.4.

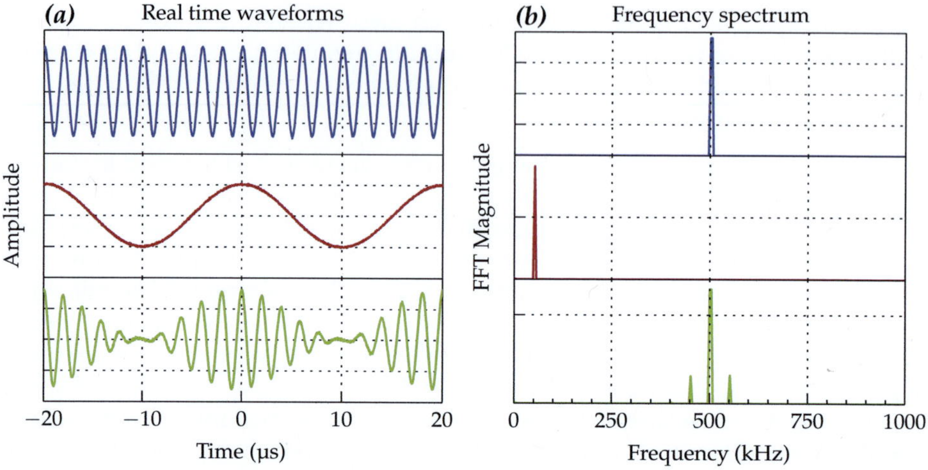

***Figure 4.6:** Real space signal ((**a**)) and corresponding frequency spectrum ((**b**)) generated by the Agilent ESG-A E4421B signal generator. (**a**) was measured with a conventional oscilloscope, (**b**) with a spectrum analyzer. For a better illustration a comparatively low carrier frequency of $\omega_{\mathrm{rf}} = 500\,\mathrm{kHz}$ and a rather high modulation frequency of $\omega_{\mathrm{mod}} = 50\,\mathrm{kHz}$ were used. The top row shows the plane, sine-like carrier wave and its corresponding frequency peak at ω_{rf}. In the middle line the low frequency modulation signal is depicted. The lowest line shows the 100% amplitude modulated signal. As expected, sidebands show up in the frequency spectrum at $\omega_{\mathrm{rf}} \pm \omega_{\mathrm{mod}}$.*

4.2.3 Generation of radio frequency voltages

Continuous wave generator

To generate radio frequency voltages, the *Agilent ESG-A E4421B* analog signal generator is used. It features frequencies from 250 kHz to 3 GHz with various modulation modes. The generator is based on a voltage-controlled oscillator (VCO) with a frequency range between 500 MHz and 1 GHz. A VCO is very similar to a transimpedance amplifier (ch. 4.1.5) but in this case the intention is that the internal operational amplifier and the feedback network of the VCO swing up. A detailed explanation is given in [90]. The frequency generated by the VCO is multiplied/divided by a constant factor to get the corresponding range. For measurement control, the generator features a serial and a GPIB interface.

Since Lock-in amplified measurements are intended, the amplitude modulation (AM) abilities of the generator are used. The modulation frequency used ω_{mod} is typically on the order of several kHz. AM is the periodic modulation of the power output or voltage amplitude at this frequency as explained earlier (ch. 3.2.1). Figure 4.6 shows a test measurement of the generator output measured with a conventional oscilloscope ***(a)***, respectively a spectrum analyzer ***(b)***. In subfigure ***(a)*** the real time waveforms are shown. They match the design waveform quite well. The frequency spectrum ***(b)*** depicts the expected lines at $\omega_{\mathrm{rf}} \pm \omega_{\mathrm{mod}}$ (green curve). The line shape can also be cross-checked (not shown). The observed peak widths are on the order of 10 Hz. This is already the limit of the spectrum analyzer as its lowest bandwidth lies within this range. The manufacturer specifies the peak width of the *Agilent ESG-A E4421B* below 1 Hz. The experimentally confirmed upper limit of 10 Hz is sufficient for the intended measurements as the linewidth should be dominated by the relaxation of the magnetic moment and therefore be much higher.

Pulse generator

For future pulsed measurements a *9530 digital delay pulse generator* from Quantum Composers Inc. was at hand. It is based on a Field Programmable Gate Array (FPGA) and allows the generation of delayed voltage pulses in the ns range. The minimum pulse width is specified with 2 ns. With this pulse generator experiments similar to the ones explained in chapter 3.1.2 should be possible. Therefore, it was programmed using its RS232 interface to produce a pump and a delayed probe pulse on two different output channels. As the generator only provides TTL/CMOS outputs with 4 V amplitude, a two channel attenuator circuit was constructed. This homebuilt circuit also features an operational amplifier to add the two channels.

Figure 4.7 shows a test measurement of the pulse generator. 50 ns wide pump and probe pulses were send through the STM setup to the nonlinear tunneling junction. The pulse shapes of the generator directly after the attenuator circuit (measured with an oscilloscope) are shown in the upper row. The pulses have an almost rectangular shape, with a rising time of about 1 ns. The cross correlation due to the nonlinear I-V characteristic is shown in the lower part of the figure. As can be seen, the shape almost fits to a triangle shape. This indicates good transmission at the corresponding frequencies. The shown type of measurement will be picked up again later for characterization of the new transmission line.

4.2.4 Custom-built measurement software

Since the measurement of magnetic resonance spectroscopy in the STM requires control and synchronization of the rf-generator, the Lock-in amplifier and the NANONIS™ electronics a custom measurement tool was implemented. The NANONIS™ software provides a LabView programming interface which can be used to write flexible, user-specific plug-ins. For magnetic resonance spectroscopy a tool similar to the default scanning tunneling spectroscopy tool was implemented. LabView is the ideal utility for this task. It possesses a design-view to create the user interface and a "block"-view to implement the program's functionality.

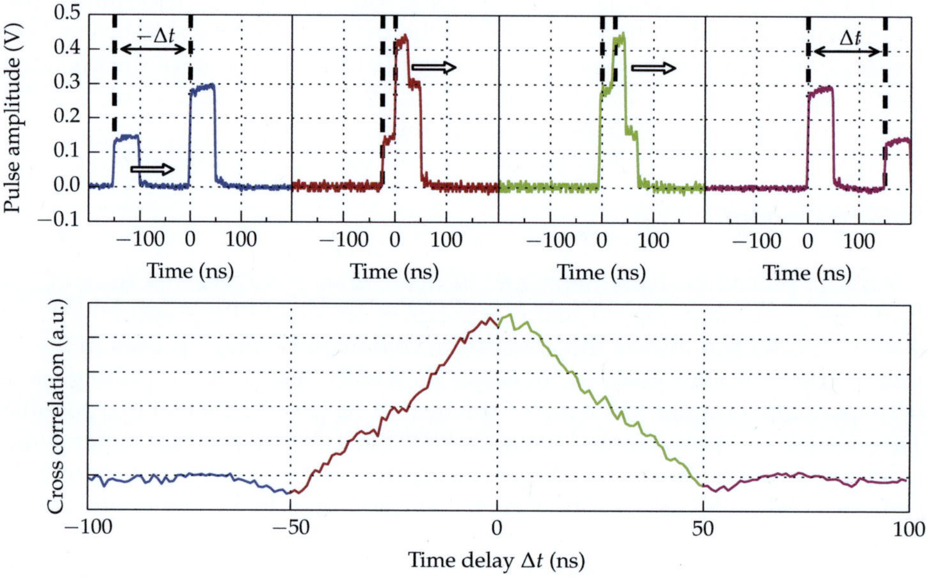

Figure 4.7: *Test measurement using the 9530 digital delay pulse generator. In the upper row different situations of the real time pulse shapes are shown. A pump pulse of* 0.3 V *and a probe pulse of* 0.15 V *were generated. Both are* 50 ns *wide. They were measured using an oscilloscope. The spectrum in the lower row shows the cross correlation of the pulses in an STM tunneling junction. Therefore, the delay time* Δt *was varied. A W-tip and a nonmagnetic Ir sample were used.*

Figure 4.8 shows the main part of the measurement tool's user interface. Various parameters can be set at initialization time to configure a magnetic resonance spectroscopy sweep. These parameters are used when a point spectroscopy measurement is started. In initial versions the entire configuration of the frequency sweep was sent to the rf-generator. This led to synchronization problems since the data acquisition is realized via the NANONIS™ electronics. Later versions use a more sophisticated approach by calculating every spectral point in advance. The points are submitted to the rf-generator one after another. Between every point the program waits a delay time according to the time constant of the Lock-in amplifier. To send every point individually provides the possibility to tune the amplitude of the radio frequency voltage at every frequency point.

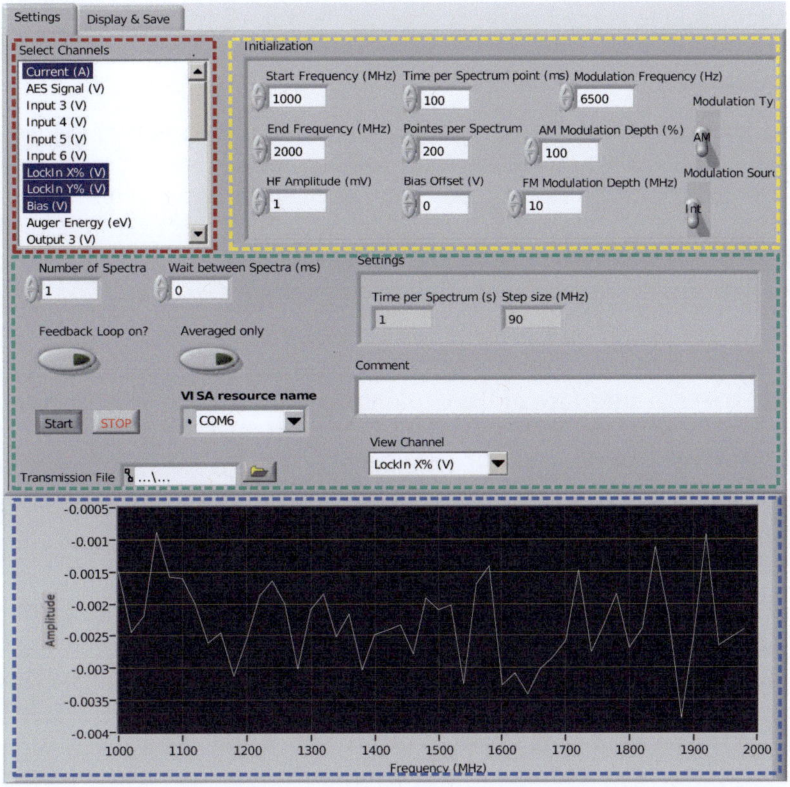

Figure 4.8: *User interface of the magnetic resonance spectroscopy measurement tool. Implemented in LabView for interconnection to the NANONIS™ scan software. In the red area the user can select the data channels that will be acquired. The yellow part shows various controls to adjust parameters of the frequency sweep like start-/ stop-frequency, radio frequency amplitude, numbers of point and many more. The main controls for a spectroscopy sweep are located in the green area. In the lower, blue part, the current measurement is shown; the desired channel can be selected. To save the acquired data, a second tab provides all relevant controls (not visible).*

This can be used to compensate for non-constant transmission effects (ch. 4.3.2). For good usability the software is able to instantly show the attained data and save or discard it upon the users decision.

Classical scanning tunneling spectroscopy measurements require to turn off the feedback loop in order to measure at constant height and not to crash during a bias sweep (ch. 2.2.2). In magnetic resonance spectroscopy this is not strictly demanded. An active feedback during spectroscopy is, for example, useful in a scenario were low frequency noise is present. To remain variable with regard to this point, the software was designed to feature both cases.

For the measurement of lateral resolved spectroscopy on a grid or a line, a secondary plug-in was developed. It accesses the configuration values that are set in the user interface of the explained main module and executes point spectroscopies at previously calculated/defined sites. All acquired data is merged into one grid measurement file for analysis and lateral representation.

4.3 Transmission measurements

Since radio frequency voltages at the tunneling junction of the STM are indispensable for the intended measurements, it is necessary to cross-check the presence of the latter. Therefore, the transmission from the rf-generator to the STM has to be measured. This equals the measurement of the actual amplitude of the radio frequency voltage U_{rf} at the tunneling junction. Usual transmission measurements are realized by connecting input and output of the sample circuit to a vector network analyzer (VNA). The VNA, in simplified terms, sends an electromagnetic wave through the circuit and measures amplitude and phase at the output of the circuit. The decrease in amplitude and the phase shift compared to the incident wave yields the sought properties. In the present setup the transmission cannot be measured in this manner as only one transmission line was embedded in the STM setup.

To get an idea of the signal shape and quality, three different methods are presented in the following sections. The first two are related to and useful for the heterodyne measurements explained earlier. The last one is based on and used for pump probe experiments.

4.3.1 Transmission measurement with a VNA

The first method to measure the transmission was carried out while the setup was disassembled. Once the vacuum chamber is vented and the STM body can be reached with arbitrary tools, it is natural to measure the transmission at least in a rough estimation. Therefore, a gauge head was connected to the STM sample stage. The rf-generator was replaced by a VNA. Its first port was connected to the start of the bias line and the second port was connected to the gauge head. In this situation the VNA allows, as mentioned above, to measure the transmission coefficient from port one to port two (S21). It should be noted that the gauge head does not exhibit $50\,\Omega$ impedance at its tip. Thus, it will additionally perturb the measured spectrum. Nevertheless, it should be possible to obtain some qualitative information about the transmission characteristics of the circuit, respectively of the new transmission line.

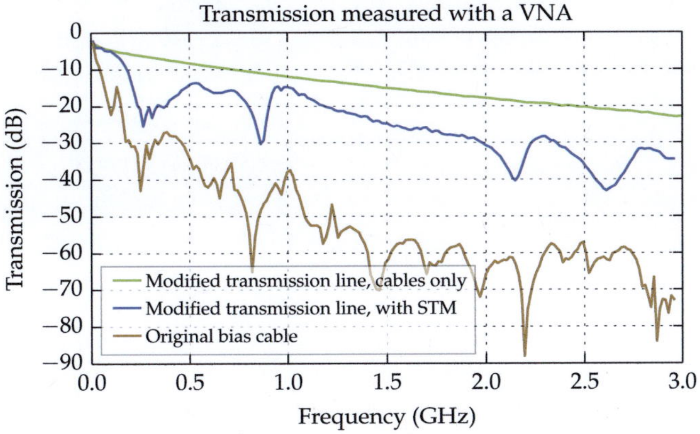

Figure 4.9: *Transmission measurements using a VNA. The curves show the transmission from port one to port two of the VNA (S21). Port one of the VNA replaces the rf-generator of the setup (fig. 4.5). The green curve illustrates a measurement where the end of the new transmission line was directly connected to the VNA's second port. A flat transmission is observed. For the other two curves a gauge head was used to connect the second port of the VNA to the STM sample stage. A clearly improved transmission can be observed, comparing the blue and the brown curve. This can be understood as many impedance mismatches were removed by the modification.*

This kind of measurements were performed before and after the installation of the new transmission line. Hence, it was possible to evaluate the achievement of the modification. Figure 4.9 shows three transmission curves. It can be seen that the transmission in the original state of the setup (brown curve) shows many sharp dips. These are resonances inside the old cables, as various impedance mismatches were present. The parts of the cable between the mismatches act as resonators with well-defined resonance frequencies. For example, in the range between 1.5 and 2.2 GHz a dip is found every 240 MHz. This corresponds to a wave length of $\lambda \approx 1.2\,\mathrm{m}$. Thus, a resonator with a length $L = n\lambda/2$ has to be present. It could be identified as the original cable inside the cryostat as this cable was approximately of this length ($L \approx 0.6\,\mathrm{m}$). This cable was not impedance matched at its ends which explains the observed resonances. Electromagnetic waves with the frequencies of the dips got adsorbed by this resonator and were not transmitted through the circuit.

__Figure 4.10:__ Rough sketch of the electrical situation inside the STM. The radio frequency input is directly connected to the sample stage. This stage is isolated from ground to allow STM operation. The tunnel junction between tip and sample stage is approximated with a capacitor. The mounting of the sample stage by piezo motors is simplified by a capacitor and an inductance.Along with the shielding of the cables of the STM tip this yields an LC-circuit.

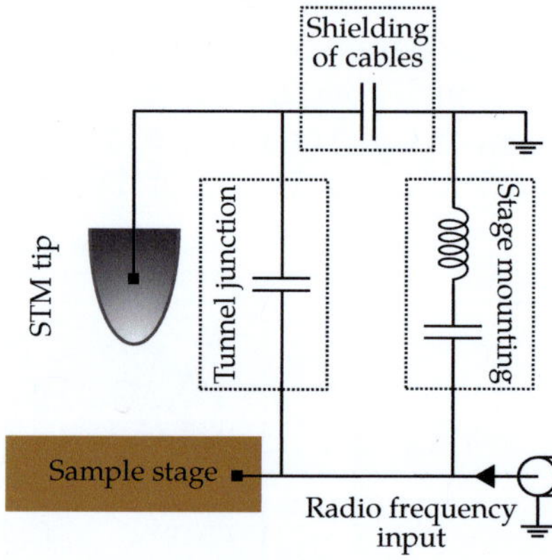

The green and the blue curve in figure 4.9 are recorded after the modification of the setup. It is obvious that much less dips are present, especially the 240 MHz periodicity is gone. To record the green curve, the end of the transmission line was not connected to the STM but directly to the second port of the VNA. A entirely flat transmission is obtained. From this curve the damping, solely coming form the new cables, can be extracted. The value of about 18 dB at 2 GHz matches the specifications of the manufacturers of approximately 19 dB (ch. 4.2.2).

In the final setup, where the transmission line is connected to the STM sample stage, the gauge head was used to measure the transmission as explained before (blue curve). It can be seen that the overall damping is slightly increased (higher slope) and four prominent dips show up. One reason for these perturbations is the way of measuring the transmission with the gauge head. As it is not impedance matched, it inevitably induces resonances and damping.

But also the electrical situation inside the STM causes the features of the blue curve. Figure 4.10 shows a rough sketch of the situation inside the STM. The tunneling junction is approximated by a capacitor between tip and sample. The sample stage, which is the end of the transmission line, is hold by piezo motors and their electrical connections. This can be simplified as a capacitor and an inductance. As also the tip has a finite capacitance to ground (due to cables), an LC-circuit is present. This can, at least qualitatively, explain the resonance features.

As a matter of course, the sample stage is isolated ($R > 1\,\mathrm{G\Omega}$) from ground for the STM to function after the modification, as well. Hence, a reflection at the end of the transmission line is inevitable. The consequential standing waves inside the whole transmission line (approximately 6 m) can be observed in measurements as explained before. An increased resolution of the blue curve in figure 4.9 yields periodic dips every 45 MHz (not shown). Along with the utilized dielectrics, this fits quite well to the cable length.

Nevertheless, the modification is a success. For example, the transmission at 1.5 GHz is increased from −60 to −25 dB. Hence, the amplitude of the transmitted electromagnetic wave is increased by a factor of 31. This raises the chances of a successful high frequency measurement in the STM setup presented.

4.3.2 Transmission measured in a tunneling experiment

The transmission measurement shown in the previous section yields sufficient evidence that a radio frequency based magnetic resonance experiment could be accomplished in the modified STM setup. Nevertheless, the transmission is far from being flat (the reasons for that were mentioned before). This means, the amplitude of the radio frequency voltage at the tunneling junction (U_{rf}) is damped with respect to the input amplitude produced at the rf-generator. This damping depends on the frequency of the oscillating voltage. Thus, the aim is to measure the actual amplitude U_{rf} for a given input voltage.

To achieve that, the signal measured by the Lock-in in a radio frequency measurement should be recalled (ch. 3.2.1):

$$I_{\mathrm{Lock-in}} = \frac{1}{2} U_{\mathrm{rf}} \frac{\partial I\left(U_0\right)}{\partial U} P_{\mathrm{T}} P_{\mathrm{S}} A \cos\left(\left(\omega_0 - \omega_{\mathrm{rf}}\right) t - \varphi\right) + \frac{1}{2} U_{\mathrm{rf}}^2 \frac{\partial^2 I\left(U_0\right)}{\partial U^2}. \quad (4.10)$$

It was shown that the signal consists of a time-dependent resonance term and a constant term. In case of a non-magnetic sample the resonance term will vanish and only the static term will remain ($I_{\mathrm{Lock-in,rf}} = \frac{1}{2} U_{\mathrm{rf}}^2 \partial^2 I / \partial U^2$). It is obvious that this term will be perturbed by transmission effects as it is proportional to U_{rf}^2. Furthermore, this term is proportional to the second derivate of the IV-curve (hence, a nonlinear IV-curve has to be present). This allows to compare the signal to a standard scanning tunneling spectroscopy (STS) measurement. In an STS experiment the Lock-in can be set to measure a signal proportional to the second derivative of the IV-curve, too ($I_{\mathrm{Lock-in,STS}} = \frac{1}{4} U_{\mathrm{mod}}^2 \partial^2 I / \partial U^2$). But since the modulation frequencies in STS experiments are in the low kHz regime, 100 % transmission can be expected. Thus, the amplitude of the modulation voltage U_{mod} at the tunneling junction is known in an STS measurement.

To unveil the amplitude of the radio frequency voltage, the two signals have to be compared: If $I_{\mathrm{Lock-in,rf}} = 2I_{\mathrm{Lock-in,STS}}$, it can be concluded that $U_{\mathrm{rf}} = U_{\mathrm{mod}}$. To achieve this, the input amplitude of a radio frequency based experiment has to be adjusted to fulfill the condition shown. The ratio of the actual amplitude achieved at the tunneling junction U_{rf} to the required input voltage produced at the rf-generator yields the transmission.

Based on the idea mentioned above, a test measurement using a non-spin-polarized tip and a non-magnetic Ir substrate was conducted. Figure 4.11*(a)* shows the initially performed STS experiment (blue curve). The Lock-in signal for a modulation frequency of $2f_{\mathrm{mod}} = 3\,\mathrm{kHz}$ and $U_{\mathrm{mod}} = 1\,\mathrm{mV}$ is plotted for a sweep of the bias voltage U_0 from -0.5 to $0.5\,\mathrm{V}$. This STS signal is proportional to the second derivative of the IV-curve as mentioned before. For comparison, the whole spectrum was already multiplied by a factor of two. As can be seen in the figure, the same curve can be obtained using a radio frequency based experiment at $f_{\mathrm{rf}} = 800\,\mathrm{MHz}$ (red curve). This confirms the theoretical deduction presented in chapter 3.2.1. To obtain the red curve, the same bias voltage and setpoint of the tunneling current as in the STS measurement were used. The important difference is the input voltage produced by the rf-generator, which had to be set to $3.13\,\mathrm{mV}$ to obtain the same signal as in the STS measurement. Thus, a transmission of about $31\,\%$ ($-10\,\mathrm{dB}$) of amplitude is present at a frequency of $800\,\mathrm{MHz}$.

To obtain a full transmission spectrum, this procedure has to be done for every point in the radio frequency spectrum. To reduce the workload and for a better frequency resolution, this was done by frequency sweeps at constant radio frequency amplitude over different parts of the spectrum. As soon as the transmission changed too drastically the amplitude was readjusted manually to gain a good signal-to-noise ratio and to probe the very same part of the IV-curve. Later, the signal was scaled by the adjusted amplitude. The required input amplitude, produced at the rf-generator, to attain a amplitude of the radio frequency voltage of $1\,\mathrm{mV}$ at the tunneling junction is shown in 4.11*(b)*. As can be seen, at low frequencies low amplitudes are sufficient since the transmission is high. At elevated frequencies the overall transmission is decreased and additional sharp peaks appear. Their origin are resonances in the circuit as discussed in the previous section. For better visibility and comparison with the transmission measurements using a VNA, the data is plotted on a decibel scale in subfigure 4.11*(c)* (green curve). A roughly linear curve is obtained, which shows a similar damping behavior as measured with the VNA. But the detailed structure of the two curves differ a lot. For example, the dips at 0.98 and $1.19\,\mathrm{GHz}$ do not match at all. Most likely the difference originates from the measurement method using the VNA (ch. 4.3.1). Only the resonance at $2.13\,\mathrm{GHz}$ seems to coincide well in both measurements.

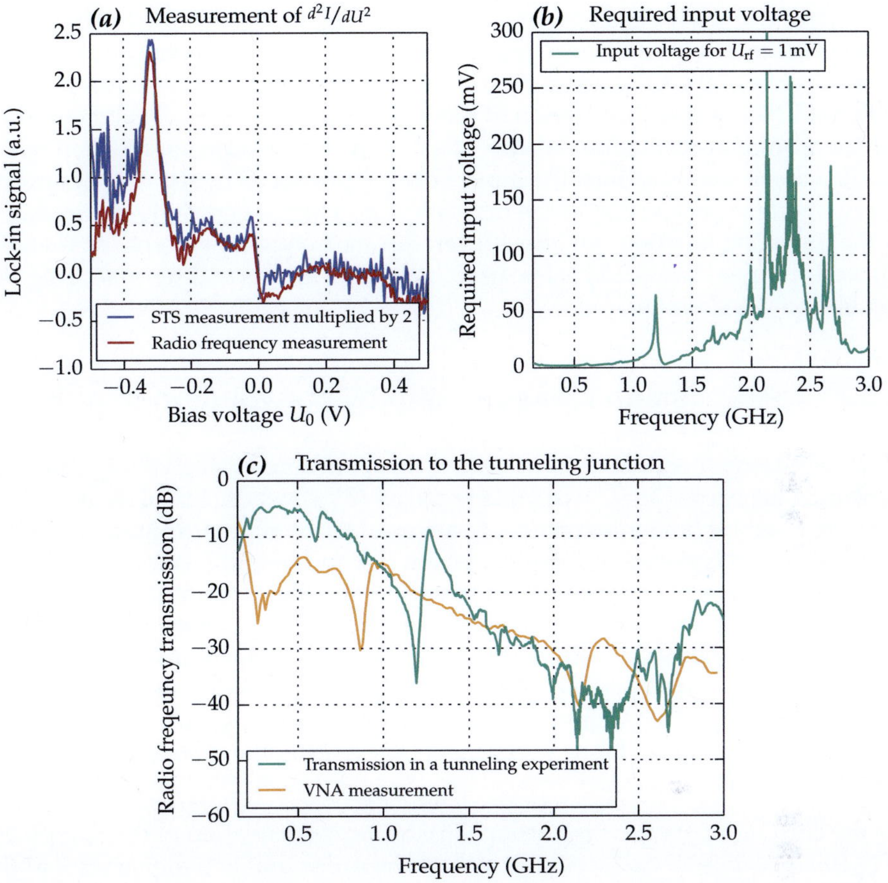

Figure 4.11: *Transmission measurement on an Ir substrate.* ***(a)****:* $2\times$ *magnified signal of an STS measurement with a modulation amplitude* $U_{mod} = 1$ mV *in blue (modulation frequency* $f_{mod} =$ 1.5 kHz*). The red curve is measured using a radio frequency measurement at* $f_{rf} = 800$ MHz*. The amplitude of the input amplitude produced at the rf-generator was set to* $U_{rf} = 3.13$ mV *to match the spectra as shown. In* ***(b)*** *the required input amplitudes of the radio frequency to gain* $U_{mod} = 1$ mV *are plotted over frequency. Part* ***(c)*** *shows the transmission obtained from* ***(b)*** *(green curve). For comparison the VNA measurement from the previous chapter is plotted (yellow curve).*

The method shown yields a very exact transmission curve, and hence is well suited to characterize the setup's performance. It allows to precisely measure the transmission in the fully assembled state of the setup, which was not possible with the VNA. A good utilization of the described measurement is the opportunity to acquire transmission data for calibration: A spectrum as shown in figure 4.11***(b)*** can be used to adjust the input voltage, produced by the rf-generator, at every frequency according to the transmission. Thus, a constant amplitude U_{rf} at the tunneling junction can be achieved during frequency sweeps. At that, the theoretical signal (ch. 3.2.1) is free of transmission effects and resonance signals can be identified easier.

4.3.3 Transmission measurement by convolution of pulses

The last measurement of the transmission was performed using voltage pulses as explained in chapter 3.1.2. Two voltage pulses ($P_{1,2}$), whose initial shapes might get perturbed by transmission effects, are send to the tunneling junction. In the present case two pulses of identical width (w) and height (A) were generated. The delay between both pulses (τ) is varied and the current increase due to the overlap of both pulses is measured. Source of the current increase is the nonlinearity of the IV-curve in the tunneling junction, which produces a current difference if the pulses overlap or not (fig. 3.3a). An exemplary spectra is shown in figure 4.12***(a)***. Two almost rectangular pulses were sent to the nonlinear IV-curve with a time delay τ. As soon as they overlap an increase in current is measured.

As the current increase is proportional to the cross-correlation of the two pulses [71], it is possible to deduce the width of the pulse that actually arrives at the tunneling junction. If the curvature of the IV-curve ($\partial^2 I / \partial U^2$) is known, the actual pulse height can be determined, as well. Therefore, the cross-correlation integral of the two identical, rectangular pulses was calculated. Its result is a triangular function:

$$I_{\mathrm{cross}}(\tau) \propto (P_1 \star P_2)(\tau) = \begin{cases} |A|^2(\tau + w) & \text{if } \tau < 0 \text{ and } \tau > -w \\ |A|^2(w - \tau) & \text{if } \tau > 0 \text{ and } \tau < w \\ 0 & \text{otherwise} \end{cases} \tag{4.11}$$

This function was fitted to the experimentally obtained spectra (fig. 4.12***(a)***). The shape is reproduced quite well, except at the edges of the triangle, i.e. the start and the end of the pulse overlap. This is coherent because the highest frequencies are present at the edge of the pulses; thus transmission effects might cause overshooting.

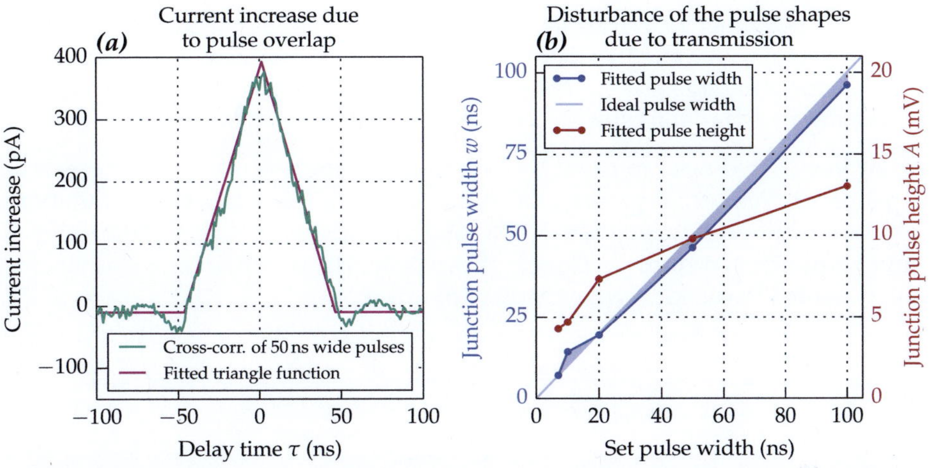

***Figure 4.12:** Transmission measurement of pulses in a nonlinear STM tunnel junction (Ir substrate). **(a)** shows the cross-correlation spectra of two* 50 ns *wide pulses (green). Their set height is unknown. A triangle function (purple) is fitted to the data, yielding a pulse width of* 46.2 ns *and a pulse height of* 9.7 mV *($\partial^2 I/\partial U^2$, which is needed for conversion, was obtained through a separate measurement of the IV-curve). **(b)**: Actual pulse width in the tunneling junction versus the set pulse width at the pulse generator (blue). The difference to the ideal case is hatched. Simultaneously the pulse height in the junction (red) is shown. It decreases continuously although the set pulse height was kept constant. Transmission effects must be present.*

The analysis of the cross-correlation was done for spectra of differently set pulse width. The obtained fit parameters of pulse widths and heights are plotted in figure 4.12*(b)*. It is clearly visible that the pulse widths in the junction corresponds quite well with the set pulse widths. This is beneficial as it should allow measurements of relaxations down to about 7 ns. This limit arises from the decrease of the produced current which vanishes in the noise background. The pulse heights tend to decrease with lower pulse width. Since that is not supposed to happen ideally, it has to be caused by transmission effects. Nevertheless, it is possible to extract the actual pulse height with the method presented. This is helpful for inelastic experiments where the actual pulse height is essential.

4.4 Summary experimental setup

In this chapter the experimental setup was outlined. As the available setup did not posses electrical wirings with sufficient performance in the radio frequency range, a modification of the setup was necessary. The core is the impedance matched transmission line from the radio frequency source to the tunneling junction. But other novelties to this setup, that were installed for this work, were presented, as well. To proof the benefit of the modifications, transmission measurements were shown. They also yield calibration data to suppress the remaining transmission effects.

According to this, a setup capable of measuring magnetic resonance signals as explained in chapter 3 is present. Of course, the setup possesses some limitations (e.g. limited frequency range from 100 MHz to 3 GHz) which restrict the range of samples that can be studied. Three different samples, which fit the requirements, are presented in the following chapters of this work.

5 Vortices in iron on tungsten

Following the theoretical background (ch. 2), the explanation of the intended measurements (ch. 3) and the depiction of the experimental setup (ch. 4), the first sample to measure magnetic resonance with an STM will be presented in this chapter. In the context of an experimental development it is consequential to wisely choose a sample which raises the possibility of a successful measurement. Such a proof of concept sample should fulfill some requirements: The resonance frequency should be in the accessible range and the system should be excitable by the presented mechanism. However, two additional properties would simplify the situation significantly: Firstly, the system should be easy to simulate theoretically in order to predict the resonance frequency and the current needed for excitation. This should simplify the measurement as not the whole frequency range has to be studied and the strength of the excitation can be estimated. Secondly, *in situ* verification of the spin polarization of the STM tip would be advantageous. This is helpful as a spin-polarized tip is necessary but often hard to achieve.

The first sample of this work, vortices in iron on tungsten (110), fulfills all these requirements. Furthermore, this model system has been widely studied in STM over the last decade. This allows to reproduce well understood samples and to use them for the intended magnetic resonance measurements.

5.1 Introduction and previous results

5.1.1 The magnetic vortex structure

A magnetic vortex is a special magnetic alignment existing in ferromagnetic particles under certain circumstances. It can be understood as a special ferromagnetic phase where the magnetization curls in a plane around a symmetry point (fig. 5.1*(a)*). This symmetry point is called the vortex core. Inside the vortex core the magnetization points out of or into the rotational plane (fig. 5.1*(b)*). This direction defines the polarization of the vortex. The first predictions of magnetic vortices date back to Bloch and Néel, who described a vortex as two intersecting domain walls [91].

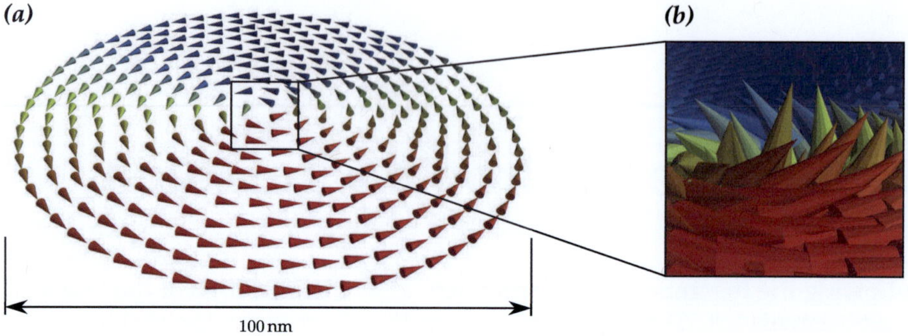

***Figure 5.1:** Illustration of a magnetic vortex. Each arrow represents the direction of the local magnetization. **(a)**: The magnetization in the depicted* 100 nm *wide disk curls around its center. In **(b)** a closer view of the vortex core is shown. It can be seen that the magnetization points out of the plane at the core position.*

***Figure 5.2:** Schematic depiction of the magnetic phase diagram for cylindrical nano-particles of different sizes. The diameter and the height are given in units of the exchange length* l_{Exc}*. Two single domain areas (red and blue) are shown for in-plane and out-of-plane magnetizations. The vortex phase is colored in green. Adapted from [93].*

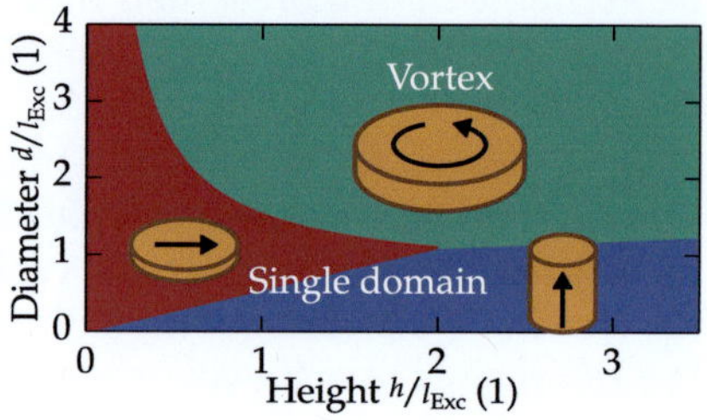

The reason for this magnetic phase to exist, is found in the counteraction between the energy of the exchange interaction (eq. 2.17) and demagnetization energy (eq. 2.23). In small ferromagnetic particles the short range exchange interaction is dominant and favors a parallel alignment between the magnetic moments; thus, a single domain state is energetically desired. As the exchange length in typical ferromagnets is only a few nanometer [92], the exchange term looses importance when particles become larger. Then the demagnetization energy, which essentially originates from the stray field and which is a long range interaction, becomes dominant. Hence, the ferromagnet develops a domain structure to minimize the stray field. However, if the ferromagnet has an intermediate size, a vortex structure can be favorable. Since the magnetization curls around the core and is parallel to all sample edges, the stray field is minimized everywhere, except at the core. Furthermore, the exchange interaction is satisfied as good as possible since almost parallel alignment is achieved throughout the ferromagnet (fig. 5.1). The magnetic phase diagram of nanometer-sized cylindrical particles is sketched in figure 5.2. The size is given in units of the exchange length.

A magnetic vortex structure formed by two intersecting domain walls was already observed in experiments many years ago, e.g. [94]. In 1999, an array of artificially patterned disks of supermalloy could be produced which exhibit a vortex ground state similar to figure 5.1 [95]. The vortex state was only observed provided that the disks exhibit an appropriated diameter-to-height ratio. The first direct observation was presented by Shinjo *et al.* in 2000 [96], who published magnetic force microscopy images of magnetic vortices in permalloy disks.

5.1.2 Dynamics of magnetic vortices

Since the intended measurements aim to yield the dynamics of magnetic moments, the interest is on the dynamic modes of the magnetic vortex structure. They can be explained by the motion of domain walls [97]. The domain walls, respectively the vortex core, can be described as a solid object which can move as a whole. This motion of the vortex core is described by the Thiele-equation [97]:

$$\mathbf{G} \times \frac{\partial \mathbf{X}}{\partial t} - \frac{\partial W(\mathbf{X})}{\partial \mathbf{X}} - D\frac{\partial \mathbf{X}}{\partial t} = 0. \tag{5.1}$$

Where $\mathbf{X}$ is the position of the vortex core. The first term is the gyro-force, proportional to the gyrovector $\mathbf{G}$, which originates from the non-uniform magnetization distribution (topological charge) [98–100]. It can be derived, as mentioned before, from domain wall motion [97, 101]. The gyrovector points along the z-direction (up or down depending on the vortex polarity) and is proportional to the saturation magnetization and the inverse of the gyromagnetic ratio. $W(\mathbf{X})$ is the potential energy which consists of exchange, demagnetization and anisotropy terms. It yields a restoring force which drives the vortex core back to its equilibrium position. The last term of the equation 5.1 is the damping. It scales with the damping constant D and is negative proportional to the velocity of the core ($\partial \mathbf{X}/\partial t$).

A gyroscopic solution can be found and was experimentally verified already in 1984 [94]. This gyroscopic mode is shown in figure 5.3, where the trajectory of the vortex core is depicted. Initially, the core was arbitrarily displaced from its equilibrium position (0,0) to the right. All three forces of equation 5.1 are depicted. In the steady-state, the gyro- and the restoring-force lead to self-oscillation of the vortex core around its equilibrium position. Omnipresent damping due to the last term of equation 5.1 demagnifies the precession and drives the core to its relaxed position.

For simple shapes of the ferromagnet, the motion can be calculated analytically [100]. This yields values of the precession frequencies of the vortex core, which could be confirmed experimentally by Brillouin light scattering on an array of many permalloy disks [102, 103]. It is obvious that the resonance frequency depends on the disk size. For permalloy disks with an aspect ratio of $0.05 < H/R < 0.4$ (H: height, R: radius), eigenfrequencies between 0.25 and 1.25 GHz are found. The analytical curve is plotted in figure 5.4. This is exactly the frequency range which is accessible in the present experimental setup (ch. 4).

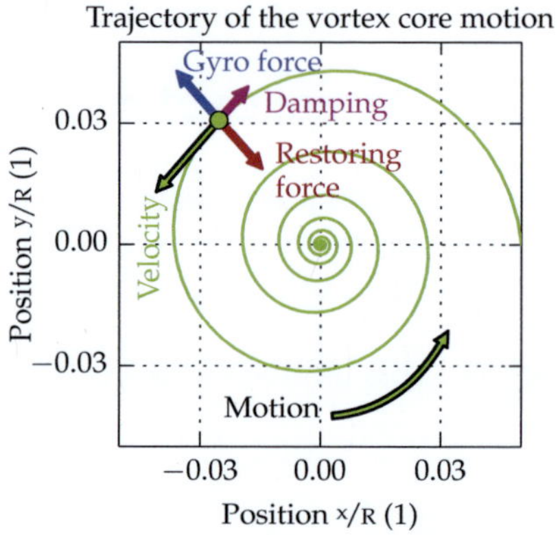

Figure 5.3: *Trajectory of the vortex core motion in units of the particle size. The vortex core precesses around the equilibrium position at (0,0). Because of energy relaxation, a damping is present which suppresses the motion and brings the vortex core at the equilibrium position to rest. The helicity is given by the polarity of the vortex core (here out-of-plane polarity is depicted). The forces of the Thiele-equation (eq. 5.1) are depicted.*

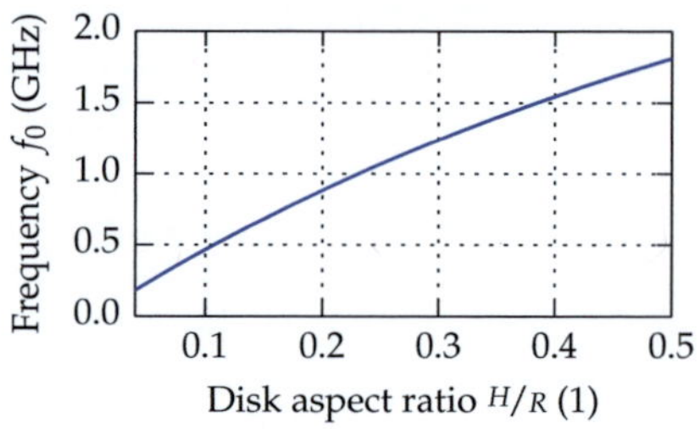

Figure 5.4: *Eigenfrequency of disk-shaped magnetic particle. The curve depicts the analytical formula of the "Two vortex model" of reference [100]. Values for the saturation magnetization and the exchange length of permalloy were used.*

5.1.3 Concept of magnetic resonance experiments applied to magnetic vortices

It was shown that magnetic vortices exhibit a gyration mode. As the eigenfrequency of this gyration mode is in the low GHz regime, the idea is to measure this dynamic mode with the concept of magnetic resonance in STM. Figure 5.5 shows a sketch of the time evolution of one gyration period. The STM tip is supposed to be spin-polarized in x-direction and is fixed above the relaxed position of the vortex core (black cross). Subfigure 5.5***(a)*** shows four snapshots of the magnetization inside a magnetic disk. A gyrating motion is shown as explained before. The color code precises the direction of the local magnetization. In part ***(b)*** of the figure the corresponding traces of the rf-voltage, the spin-polarized conductance and the resulting current are plotted. The resonant case, a phase lag of zero and zero bias voltage is assumed.

In the time ranges I and V the magnetization at the black cross is parallel to the direction of the tip (red color code), thus a high conductance is present ($+G_{\mathrm{TMR}}$). Along with the rf-voltage being in the high part of the phase a current flows (I_{rf}). In range II and IV the magnetization is 90° oriented with respect to the tip (green color code). This yields an intermediate conductance, but due to zero voltage no additional current is created. Range III is the counterpart to range I and V. An antiparallel alignment is present (blue color code). Since the rf-voltage is in the negative section of the phase, again a current is produced. By completing the full gyration sequence the current is given by heterodyne mixing: A sine curve at twice the original frequency and with a constant shift of I_{het} (dark green) is present. The oscillating component of the current is not detectable by the IV-amplifier as it is too fast, but the offset I_{het}, which can also be understood as the time average, is a constant and can be measured as explained earlier (ch. 3).

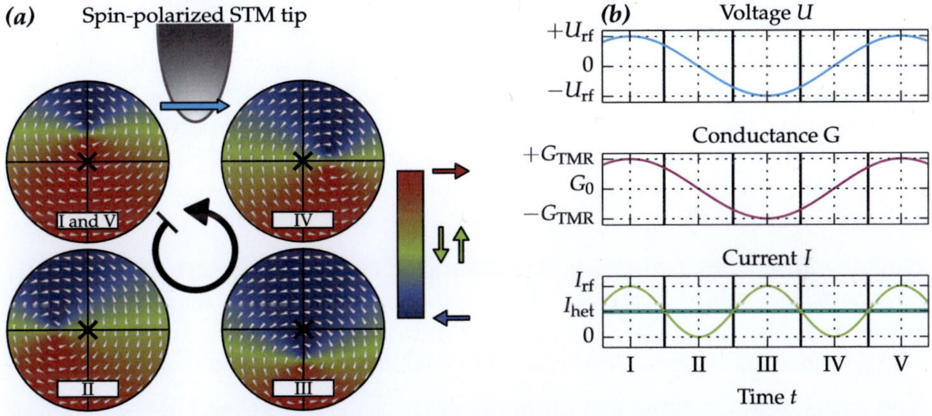

Figure 5.5: Time sequence of a resonance measurement applied to magnetic vortices. In ***(a)*** *four snapshots of the local magnetization of a gyrating vortex are shown. The color code shows the alignment with respect to the tip spin polarization. In* ***(b)*** *the corresponding time traces of bias voltage, conductance and current are plotted. The resonant case and no phase lag is assumed. This yields a heterodyne current (dark green) as explained in the text.*

5.1.4 Iron on tungsten (110)

Unfortunately, the mentioned permalloy disks (ch. 5.1.2) can only be produced by electron beam lithography. This technique is not available in the present experimental setup. And since SP-STM is a very surface sensitive technique, the samples have to be prepared *in situ* by self-organized growth to achieve sufficient cleanliness. Thus, such disks cannot be produced and a similar, realizable system had to be found.

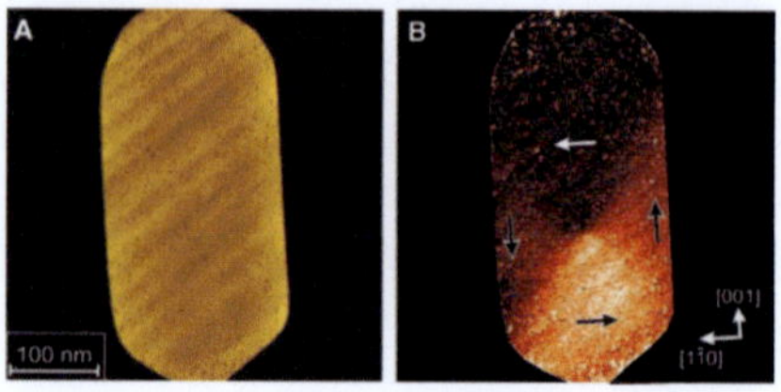

Figure 5.6: ***(a)*** *topography and* ***(b)*** *map of the* $^{dI}/_{dU}$ *signal of a single* 8 nm*-high Fe island recorded with a Cr-coated W tip. The vortex domain pattern can be recognized in* ***(b)****. Arrows illustrate the orientation of the domains. From [54]. Reprinted with permission from AAAS.*

A prominent system is given by iron islands on tungsten (110). This is the first system where the magnetic vortex structure could be confirmed by means of STM [54]. Iron islands were prepared by self-organized growth on a W(110) substrate and were studied with a Cr-coated W tip. Figure 5.6***(a)*** shows the topography image of an Fe island of about $200 \times 500 \times 8\,\mathrm{nm}$. The simultaneously recorded conductance map ($^{dI}/_{dU}$-map) yields the magnetic contrast (fig. 5.6***(b)***). As an in-plane polarized tip was used, a contrast between different in-plane orientations is expected. This is achieved very well. It can be seen that the contrast varies from bright in the parallel aligned region to dark in the antiparallel aligned area. The $90°$ aligned areas show an intermediate contrast. Thus, it was confirmed that the magnetization curls around the vortex core center. The out-of-plane magnetization could be confirmed by an out-of-plane magnetized STM tip (not shown).

The transition from a single domain island to the vortex structure is naturally thickness-dependent. The islands have to be at least 6 nm high to possess a vortex ground state [104]. This phase boundary between single domain and vortex state could also be confirmed in Fe islands on the W(100) surface [55].

As the islands are so large that quantum effects can be neglected, micromagnetic simulations provide a decent, widely confirmed approach to simulate the magnetism of the system. The background of these simulations will be explained later (ch. 5.2.2); however, it is noteworthy that the dynamics of the system can be unveiled by the latter. This is helpful as, to the knowledge of the author, no experiments on the dynamics of the system Fe on W(110) have been published yet. Nevertheless, a dynamic behavior, similar as explained before (ch. 5.1.2) can be expected and the simulations will confirm this assumption.

Therefore, the presented Fe islands are an adequate system to test the concept of a magnetic resonance measurement in an STM as explained in chapter 3. All requirements, like the confirmation of the spin polarization of the tip, the feasibility to easily simulate the system and a resonance frequency in the reachable range, are met. Even the strength of the excitation should be sufficient, as will be shown later. Thus, it is self-evident to reproduce this well-understood system and use it as a proof of concept system.

It is noteworthy that the measurement of the vortex precession does not fulfill the aim of measuring a single magnetic moment. Only the dynamics of a continuum magnetization are targeted. But as this is an experimental development an easy, well-understood system is preferable as a proof of concept.

5.2 Experimental realization

5.2.1 Preparation and growth

The first step to reproduce the vortex structure in Fe islands, as explained before, is the preparation of the sample. Since the islands have to be grown on a W(110) crystal, the latter has to be cleaned first. This is done by glowing the crystal in O_2-atmosphere at a pressure of about 5.0×10^{-7} mbar (this was already mentioned in chapter 4.1.3). Through this treatment, carbon, the main impurity, segregates to the crystal surface and react with O_2 to form CO and CO_2. Subsequent, short, high-temperature flashes up to 2200 °C remove the remaining O atoms and yield an atomically clean W surface.

Figure 5.7 shows three STM topography images. In subfigure ***(a)*** an early stage of the cleaning process can be seen. The black dots and areas are most likely C atoms or clusters. In the image shown about 30 % of the surface area is covered with adsorbates. After several cleaning cycles, as explained before, the second image (fig. 5.7***(b)***) could be recorded. As the lateral size of both images is equal, it is obvious that a significant reduction of adsorbates could be achieved. Here, only about 4 % of the surface area is covered with foreign atoms. This purity is sufficient for the growth of Fe islands. The third image (fig. 5.7***(c)***) shows a larger area of the W crystal. After the cleaning process mono-atomic steps can be observed. A stepped substrate is essential for a proper growth of the Fe islands [105, 106].

After the cleaning process, Fe was epitaxially grown on the W surface. The utilized metal evaporator was explained in chapter 4.1.3. The growth of this heteroepitaxial structure was extensively studied in the past [104, 105, 107–109]. In the first atomic layers the growth is dominated by the difference in surface energy between Fe and W. Although there is a significant lattice mismatch between Fe (next neighbor distance: 247 pm) and W (next neighbor distance: 273 pm) of about 10 %, the higher surface energy of W enforces layer-by-layer growth in the first atomic layers. Thus, the Fe lattice is stretched to match the W one.

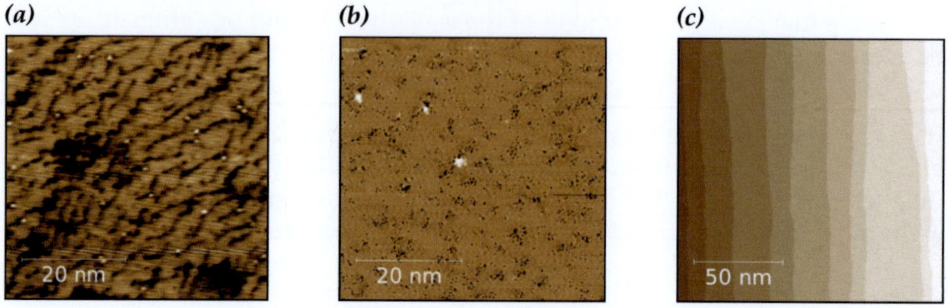

Figure 5.7: *STM topography images of the W(110) crystal surface.* ***(a)****: Early stage of the cleaning process. The black areas are impurities on the W surface.* ***(b)****: Reduced contamination through cleaning as described in the text.* ***(c)****: Lager image which shows clean, atomically stepped W(110) surface. Tunneling parameters:* $I_T = 1\,\mathrm{nA}$, $U_0 = 1\,\mathrm{V}$

If more than one atomic layer of Fe is deposited on the W substrate, the temperature of the latter influences the growth significantly. The deposition at a substrate temperature of $T_{\mathrm{W(110)}} \approx 100\,°\mathrm{C}$ and amounts of more than 4 atomic layers results in Fe strips along the crystallographic [001] axis of the substrate (fig. 5.8*(a)*). For an island growth analogous to reference [54], an annealing treatment after the deposition is necessary. Heating of the system to about $575\,°\mathrm{C}$ for about 10 to $15\,\mathrm{min}$ is sufficient. In figure 5.8 three different stages of the annealing treatment are shown. ***(a)*** shows the initial state as explained before. In ***(b)*** an intermediate state is presented. There, the sample was annealed for only $3\,\mathrm{min}$ at a temperature of only $500\,°\mathrm{C}$. It can be seen that islands start to built up and the stripe density decreases. This can be understood in terms of "Ostwald ripening" as material converges. It should be mentioned that one mono-layer of Fe remains intact on the substrate. The final Stranski-Kranstanov growth is shown in subfigure ***(c)***. Here, the sample was held at $585\,°\mathrm{C}$ for more then $10\,\mathrm{min}$. The growth can be explained as the thermal energy is sufficient for the Fe atoms to overcome various diffusion barriers (e.g. diffusion over step-edges). Thus, they can move to the thermodynamical preferred state, which is, as mentioned before, given by Stranski-Kranstanov growth as a big lattice mismatch is present.

It should be mentioned that the entire growth process can yield various island shapes. This is caused by the many parameters that influence the growth process: Deposition amount, substrate temperature during deposition, post deposition annealing temperature and duration, step density and cleanliness of the crystal; altogether they offer a wide parameter space with various, slightly different resulting sample structures.

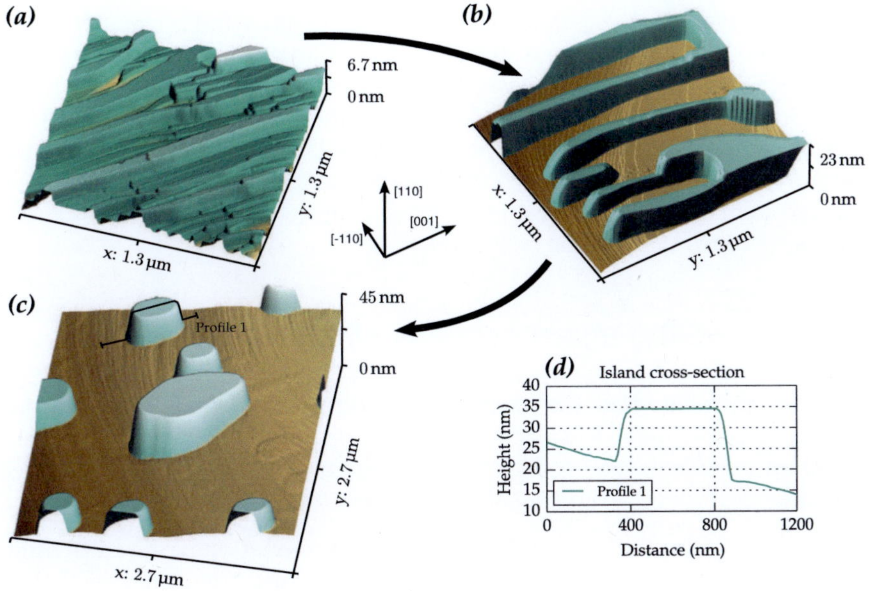

Figure 5.8: *3D illustration of STM topography images showing the growth of Fe on the W(110) surface.* ***(a)*** *shows the situation directly after the deposition of Fe at* $T_{W(110)} \approx 100\,°C$*. Fe is shaded in green. In* ***(b)*** *a scan of a partially annealed sample is shown (*$t_{anneal} = 3\,min$*,* $T_{W(110)} \approx 500\,°C$*). Picture* ***(c)*** *is an example of well-grown islands through post deposition annealing for more than* 10 min *at* 575 °C*. Perfect three dimensional islands with sizes inside the magnetic vortex phase are achieved. In* ***(b)*** *and* ***(c)*** *Fe is again shaded in green, except the epitaxial wetting layer. In subfigure* ***(d)*** *the cross-section of an island in* ***(c)*** *is plotted. An island height between* 10 *and* 17 nm *is achieved.*

The shown, well-grown islands exhibit an atomically flat surface in the crystallographic [110] direction. A cross-section of an exemplary island is shown in figure 5.8*(d)*. In the ideal case they are between 8 and 17 nm high, about 400 nm long and about 250 nm wide (aspect ratio between 1.7 and 1.5). Thus, the islands posses the correct shape to be located in the magnetic vortex phase. Therefore, the next step is to confirm the magnetic structure. To affirm the data obtained in magnetic measurements, micromagnetic simulations will be used and briefly introduced in the next section.

5.2.2 The utility: micromagnetic simulations

As previously mentioned, the magnetic structure of the islands can be simulated utilizing micromagnetic simulations. This is possible since the islands are so big that quantum effects can be neglected and the magnetism is well-described by the classical Landau-Lifschitz-Gilbert equation (eq. 2.18).

Micromagnetic simulations are based on the discretization of the LLG equation in space and time. The actual derivation is beyond the scope of this work and can be found elsewhere [110]. But it is worth mentioning that the sample geometry has to be divided into small elements. Two approaches exist: the finite differences method, where the geometry is divided into a regular lattice, and the finite elements method, where the geometry is represented by many tetrahedrons. Finite differences have the advantage that the energy term from the demagnetization field can be calculated by Fast-Fourier-Transform (FFT), which yields a gain in performance. The advantage of using finite elements is the possibility to model complex shapes more accurately.

Within the scope of this work, the package *nmag* from the University of Southampton was used [111]. It is based on the finite element approach and features a handy python interface. The package handles the whole calculation of the magnetism. The user only has to supply the mesh of the geometry and the material parameters. These are the saturation magnetization, the exchange constant, magnetocrystalline anisotropies and the LLG damping factor. Thereby, the package allows to calculate the time development of the system by numerical integration of the LLG equation at every lattice point. The interface allows individual time steps or a full relaxation run, to a (local) energetic minimum. It is noteworthy that the package also allows to add an external magnetic field term and a spin-torque term to the LLG equation (eq. 3.15). These fields can be defined globally or at individual lattice points of the geometry mesh. This feature is useful to simulate the excitation of the system according to chapter 3.2.2.

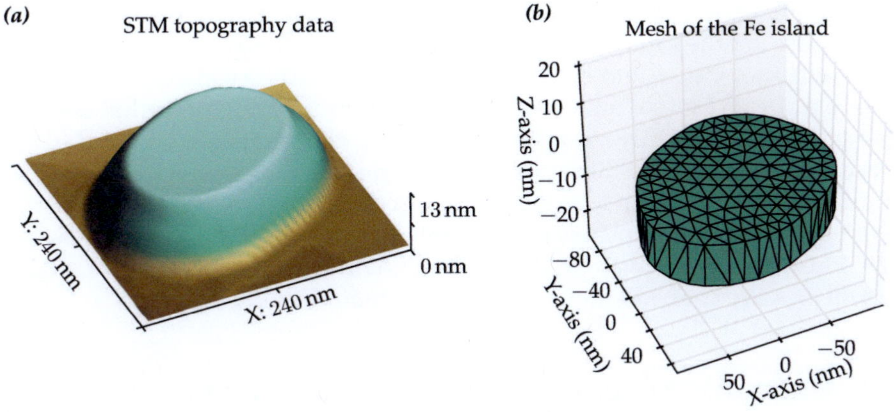

Figure 5.9: Illustration of the meshing of an exemplary Fe island for micromagnetic simulations. In ***(a)*** *the 3D representation of an STM topography scan of an island is shown. The island is shaded in green. The 3D plot in subfigure* ***(b)*** *shows the meshed geometry of the island in* ***(a)****. The maximum edge length of the tetrahedrons was set to* 15 nm*. This yields* 305 *mesh points which are connected to form* 829 *tetrahedrons.*

The first step to simulate an Fe island, as shown before, is to model the island shape based on actual measured STM topography data. Therefore, a short script which detects constant height contours in the image was developed. These contours were simplified and replaced by planes to define the island geometry. In the next step, the geometry was meshed using the program *netgen* [112]. An example is shown in figure 5.9. The Fe island in ***(a)*** was approximated by the finite element mesh in ***(b)***. For illustration the maximum length of a tetrahedron edge was set to be 15 nm. This is far too long as the cell length should be below the exchange length [113]. A small cell size is necessary as the approximation in the simulation is only valid for small angles between the magnetizations of neighboring lattice points [114, 115]. Hence, to validate a simulation it is common to verify that the maximum angle between two cells is below 30°. It is natural to increase the number of cells to achieve a more stable simulation, but this number is of course limited by computational power and especially memory consumption. A computer system consisting of an *Intel® Core™ i7-5820K* processor with 6 cores and 32 GB RAM was used for the simulations. It allows geometries up to approximately 400 000 unit cells to be calculated at reasonable times (some days of calculation), as the package allows subdivision of the geometry and distribution of the calculations to different CPU cores (parallelization). This computational power is sufficient for reasonable calculations of the magnetism of the Fe islands.

The only additional parameters for a simulation are the material parameters mentioned. These are the saturation magnetization, the exchange interaction energy and the anisotropy type and constants. As the Fe islands are many hundreds of nm large it is valid to use bulk values of the saturation magnetization and the exchange interaction. Only the effective anisotropy constant was adjusted to the thickness of the islands and the crystallographic axis was aligned according to literature values [104, 116–118]. Thus, all necessary requirements for realistic simulations of the Fe island are met. As micromagnetic simulations are based on the LLG equation, which describes the dynamic behavior of ferromagnets, they will provide confirmation of static and dynamic measurements.

5.2.3 Magnetism - spin-polarized measurements

To resolve the magnetic structure of the Fe islands, shown in section 5.2.1, spin-polarized measurements are needed. Therefore, a spin-polarized STM tip was prepared (ch. 4.1.3).

Figure 5.10 shows the result using an in-plane-polarized tip. This was achieved by coating of a W tip with Fe. In subfigure ***(a)*** a 3D representation of the geometry of an island is shown. The simultaneously recorded conductance map at a bias voltage of $U_0 = -400\,\mathrm{mV}$ was used as color-coding. At this voltage a strong magnetic contrast, in accordance with [104], was found. The red color represents a high conductance and thereby a parallel alignment to the tip polarization. In the blue area a lower conductance was found which denotes antiparallel alignment. The green parts exhibit intermediate conductance which indicates $90°$ rotation with respect to the tip. In 5.10***(b)*** the topview of the magnetic contrast is shown. It matches the micromagnetic simulations in subfigure ***(c)*** quite well. It is important to note that the polarity of the vortex cannot be determined as the polarization of the tip is twice degenerate. In the simulation an out-of-plane polarity was assumed which implies a tip polarization to the right.

As both electrodes, tip and sample, consist of the same material (Fe) the data can be used to estimate the magnitude of the spin polarization. A line section around the vortex core is shown in subfigure 5.10***(d)***. A fit of a cosine function yields a TMR (ch. 2.2.3) of about $TMR = 17\,\%$. Thereby, a spin polarization of about $P_{\mathrm{Fe}} = 0.31$ is present. This is close to literature values of Fe ($P_{\mathrm{Fe,lit}} = 0.42$) [46].

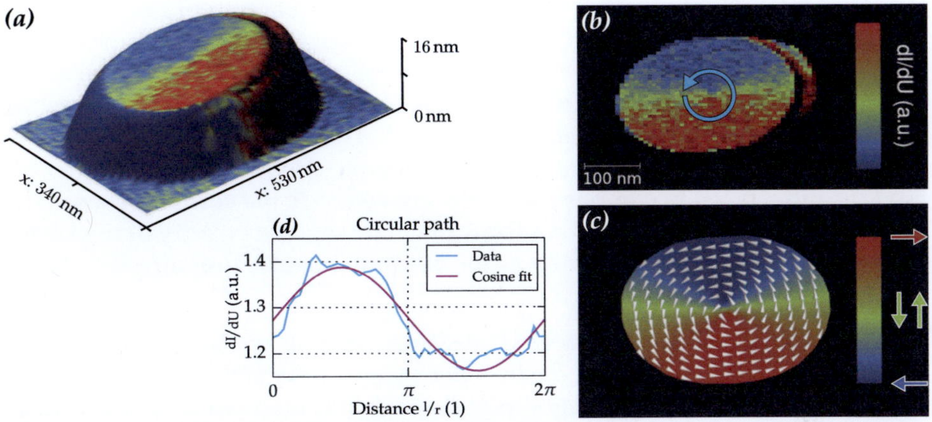

Figure 5.10: *Spin-polarized measurement of an Fe island using an in-plane-polarized STM tip. A W tip was covered with Fe which yields in-plane magnetic contrast.* ***(a)*** *shows a 3D representation of the topography data of the island. The magnetic contrast was overlaid. The red color shows high, the green one shows indermediate and the blue one shows low conductance. The data was acquired at* $U_0 = -400\,\mathrm{mV}$ *and* $I_0 = 1\,\mathrm{nA}$. ***(b)*** *depicts the magnetic contrast in topview for comparison to the micromagnetic simulation* ***(c)****. The white arrows in* ***(c)*** *are probes of the simulation data showing the direction of the magnetization. The colored surface represents the in-plane contrast at the island surface (linear interpolated). A tip spin polarization to the right was assumed. Subfigure* ***(d)*** *shows the circular cross-section of the magnetic data in* ***(b)****. The data was extracted in a radius of* 45 nm *around the vortex core. The course matches a cosine shape.*

Figure 5.11: *Out-of-plane magnetic contrast on an Fe island utilizing a Cr coated W tip.* ***(a)*** *shows a measurement and* ***(b)*** *the corresponding micromagnetic simulation. The increase of the local conductance inside the vortex core yields a parallel alignment of tip and sample spin polarizations. Outside the core no contrast is found. The simulation and the data match very well. Because the sign of the spin polarization of the tip is unknown the polarity of the vortex in* ***(b)*** *could also be reversed. The data in* ***(a)*** *was acquired at* $U_0 = -400\,\mathrm{mV}$ *and* $I_0 = 1\,\mathrm{nA}$.

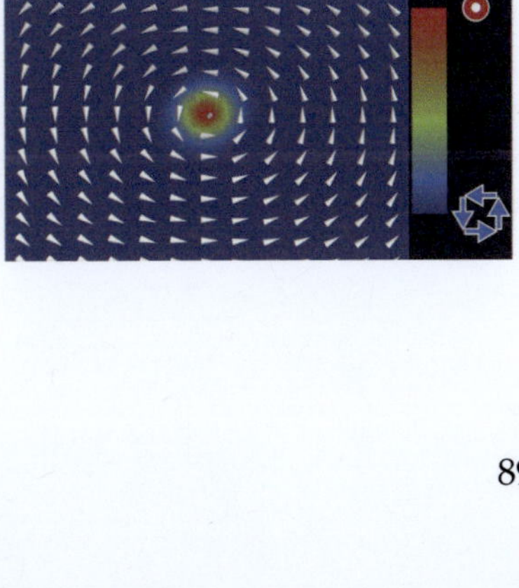

In the scope of spin-polarized measurements also the out-of-plane contrast could be verified. Therefore, a W tip was covered with approximately 40 ML of antiferromagnetic Cr. This yields an out-of-plane sensitive tip [75]. An image of a vortex core is shown in figure 5.11***(a)***. A bright core with a diameter of about 20 nm was found. Outside the core, where the magnetization curls in-plane around the core, no contrast can be observed. The core appears bright as the local magnetization points in (or out of) the surface plane. Hence, a higher (or lower) conductance in comparison to the area outside the core is present. The simulation in subfigure 5.11***(b)*** confirms the findings. The polarity of the vortex core can again not be determined as the absolute polarization direction of the tip is unknown.

Through the shown spin-polarized measurements it can be concluded that the grown islands exhibit the desired magnetic vortex structure and thereby are suited for the intended magnetic resonance measurements. This magnetic ground state of the island is the perfect starting point for dynamic simulations and measurements which will be shown in the next sections. It is noteworthy that the shown magnetic measurements not only proof the vortex structure of the islands but also confirm the spin polarization of the STM tip which is necessary for the measurements of the dynamics.

5.2.4 Simulations of the dynamics

After unveiling the static magnetic structure in the previous section, it is obvious to process towards the measurement of the dynamic of the vortex (ch. 5.1.2). Therefore, again micromagnetic simulations were used to appraise the dynamical properties of the islands under study.

Excitation by a constant external field

The objective of dynamic simulations is to estimate the resonance frequency and modes of a given island. Therefore, the relaxed state, shown in the previous section, has to be excited. In the simulations this can be achieved by a simple, static, external magnetic field, e.g. in the sample plane. This is shown in figure 5.12***(a)***. It can be seen that the magnetic field of about 10 mT in x-direction displaces the vortex core about 12 nm in y-direction. This can be understood as the red area, where the magnetization of the island is parallel to the external field, is energetically favored (Zeeman-energy) and therefore increased. Thus, the vortex core is displaced upwards out of its equilibrium position and thereby excited.

To study the dynamics, the external field is suddenly turned off at $t = 0$ (green curve in figure 5.12*(c)*) and the time evolution of the magnetization is calculated. Figure 5.12*(b)* shows the state of the island every 100 ps. It can be seen that the vortex core precesses around its relax position (black cross) as expected according to the findings in permalloy disks by other groups (ch. 5.1.2). The graph in figure 5.12*(c)* (purple curve) shows the normalized component of the magnetization in x-direction at the relax position. It can be seen that a fine sinusoidal oscillation is achieved. A rather small LLG damping factor of 0.01 accounts for the decay of the oscillation over several periods. This well defined gyration mode is the target of the magnetic resonance measurements.

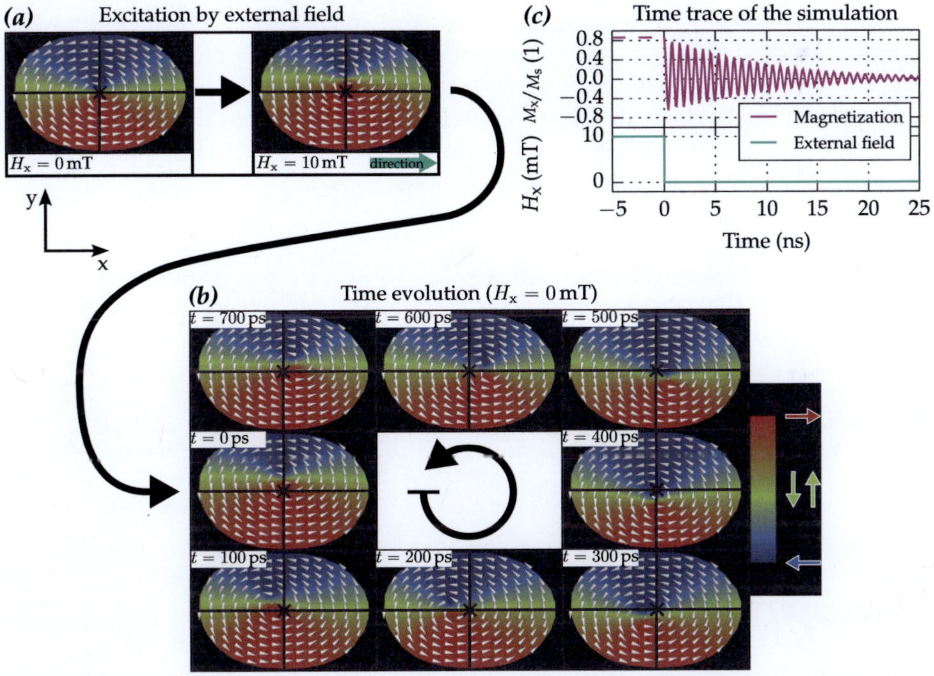

Figure 5.12: *Dynamic simulation of an Fe island as shown before. The excitation via an external magnetic field in x-direction is illustrated in* ***(a)****. The vortex core is displaced in y-direction. At* $t = 0$ ps *the external field is turned off.* ***(b)*** *shows the time evolution of the magnetization of the island. The vortex core precesses around its equilibrium position (black cross). The magnetization in x-direction was extracted over time at the equilibrium position. It is plotted in subfigure* ***(c)****. A fine sinusoidal curve is obtained which decays exponentially due to the damping term in the LLG equation.*

The frequency of the gyration mode and thereby the scan range of the experimental measurements can be obtained by Fourier-transformation of the time-trace shown in figure 5.12*(c)*. The Fourier spectrum is plotted in figure 5.13 (red curve). A clear peak at the resonance frequency of $f_0 = 1.31\,\mathrm{GHz}$ can be seen.

Similar to magnetic resonance in single domain ferromagnets an external field will shift the resonance frequency due to Zeeman-interaction (ch. 2.1.2). This can be utilized in the measurements and the simulations to verify the magnetic origin of the resonance. The simplest behavior is expected for a field out of the sample plane. Fields in this direction will not influence the vortex core's relax position and at magnitudes of some hundred mT the shape will not be influenced either as this is much smaller than the exchange and demagnetization fields. However, the resonance frequency of the gyration mode will be influenced proportional to the corresponding gyromagnetic ratio ($f_0 = \gamma B_z$). In figure 5.13 the resonance spectra for three different magnetic fields are shown. It can be seen that the external field, as it is parallel to the polarity of the vortex core, increases the resonance frequency. On the contrary, a magnetic field opposite to the core polarity will lower the resonance frequency. A linear fit (not shown) of several simulations with different external fields in z-direction yields a gyromagnetic ratio of $\gamma = 664\,\mathrm{MHz/T}$ for the examined island shown.

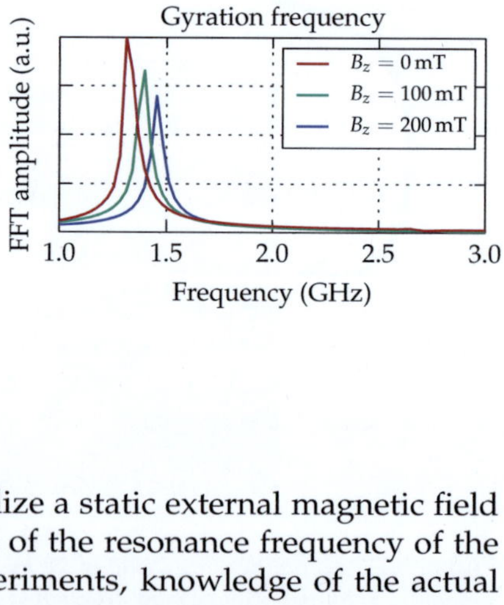

Figure 5.13: *Fourier spectra of the time evolution of the magnetization in x-direction according to figure 5.12**(c)**. A clear resonance (gyration mode) is found at $f_0 = 1.31\,\mathrm{GHz}$ (red curve). An external field in z-direction shifts the resonance. Due to parallel alignment between the field and the vortex polarity the frequency increases with* $664\,\mathrm{MHz/T}$.

Excitation by spin-torque

The shown, simple simulations, which utilize a static external magnetic field for excitation, are sufficient to gain values of the resonance frequency of the simulated island. For comparison to experiments, knowledge of the actual oscillation amplitude of the magnetization below the tip and the phase difference to the excitation current, respectively the rf-voltage, would be advantageous (eq. 3.13). This can only be achieved by correct modeling of the dominant excitation mechanism. It was shown in chapter 3.2.2 that the spin-torque effect is dominant in the present STM geometry. It is convenient that the utilized *nmag* package allows to include the spin-torque term according to the Zhang-Li model [81].

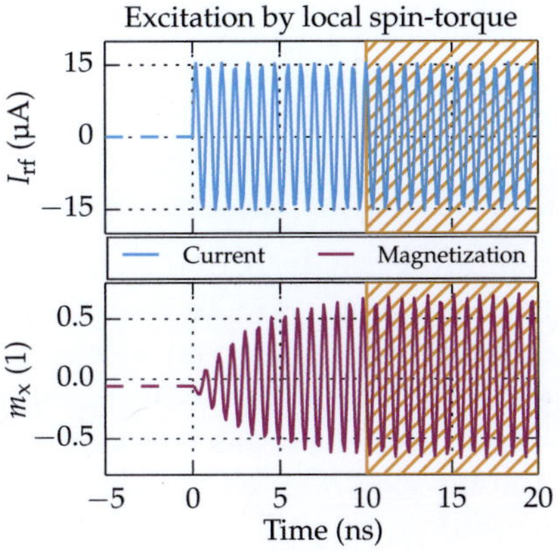

Figure 5.14: *Simulation of an Fe island utilizing spin-torque excitation. In blue, the specified current (polarized in x-direction) is plotted. A frequency $f_{rf} = 1.31\,\text{GHz}$ and an amplitude $I_{rf} = 15.5\,\mu\text{A}$ were used. The purple curve shows the x-component of the magnetization (normalized to M_s) given by the simulation. Both curves show the local behavior at the tip position. In the orange areas the curves were fitted with a cosine function to obtain values of amplitude and phase.*

This enables to accurately model the excitation by a current flowing between the STM tip and the island. It is noteworthy that the spin torque term can be specified individually at every lattice point of the simulation. This allows to simulate the local injection of the current below the tip.

Simulations of the same island as in the previous section were conducted. Starting from the relaxed state of the island (fig. 5.10) the area of the spin-polarized current was defined as a sphere (radius $r = 1\,\text{nm}$) located at the vortex core position. The direction of the spin-polarized current was chosen to be along the x-direction, which is the obvious choice when a in-plane STM tip is employed.

Under these conditions simulations up to $t = 50\,\text{ns}$ were performed. At every time step of the simulation ($100\,\text{ps}$) the magnitude of the spin-polarized current was adjusted to follow a sinusoidal curve over time with a frequency f_{rf} and an amplitude I_{rf}. This condition is a very good approximation of the situation in an STM with an applied rf-voltage as explained in the experiments design (ch. 3).

A single time-run of the simulation is shown in figure 5.14. The initial value of $f_{rf} = 1.31\,\text{GHz}$ was taken from the basic simulations using field excitation, which were presented in the previous section. The initial amplitude $I_{rf} = 15.5\,\mu\text{A}$ was chosen to be quite large (in terms of an STM) to yield a strong signal. The blue curve shows the time trace of the spin-polarized current. A spin-polarization of the tip, respectively of the current, in x-direction was arbitrarily chosen. The current follows, as defined, a sinusoidal curve. The normalized magnetization component in x-direction is plotted in purple (this corresponds to the projection of the sample spin-polarization onto the tip spin-polarization). At $t < 0\,\text{ns}$, m_x is almost zero. This is natural as the vortex core is probed which has an out-of-plane magnetization. With the initiation of the simulation the magnetization starts to oscillate, following the frequency of the driving current. After (in this case) about 10 ns the transient effect is over and a constant oscillation is accomplished. This can be understood as the homogeneous oscillation is damped out and the behavior is solely determined by the excitation. It must be emphasized that thereby the resonance case (ch. 3) is implicitly present.

For proper analysis of the simulation both curves of figure 5.14 were fitted in the areas marked in orange (when steady state oscillations are obtained) with a cosine function ($m_{\mathrm{x}}(t) = m_{\mathrm{x},0} + A\cos\left[2\pi f_{\mathrm{rf}} t + \varphi\right]$) . It is valid to discard the transient effect when comparing to experiments. There, integration times of hundreds of µs to ms will be used; hence, a static oscillation can be assumed. The fits yield the amplitude, the frequency and the phase difference between excitation and oscillation. These insights are very valuable for estimation of the expected resonance signal.

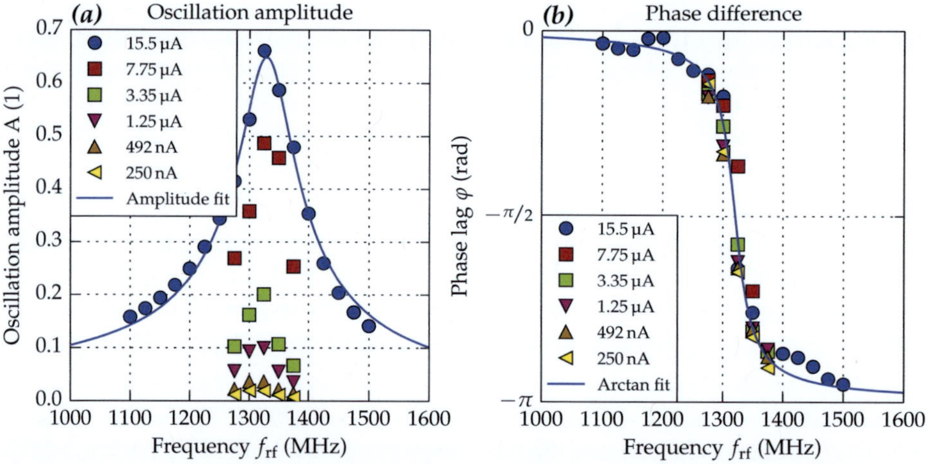

Figure 5.15: (a) amplitude and (b) phase difference obtained from many individual simulations using spin-torque excitation. All data points were obtained through simulations shown in figure 5.14. The frequency range between 1100 and 1500 MHz was studied. The amplitude of the excitation current (depicted by the different symbols) was gradually decreased to reach values that are accessible in an STM. The amplitude follows a Lorentzian lineshape and the phase difference shows an arc tangents behavior. Both dependencies could be fitted to obtain an analytical form.

To simulate an experimental frequency sweep, the shown procedure was repeated for a list of several excitation frequencies f_{rf}. Naturally different, gradually decreased values for the excitation amplitude I_{rf} were used as well. Figure 5.15 shows the values achieved by fitting each dataset as explained. In 5.15***(a)*** the oscillation amplitude can be seen. A fine Lorentzian lineshape was found which is centered at the resonance frequency. The amplitude decreases for lowered excitation currents which is expected as the driving torque is not sufficient to overcome the restoring force inside the Fe island anymore. Realistic, in STM attainable rf-currents below 1 µA are also shown in the graph. An excitation current of about 250 nA tilts the direction of the magnetization in the island approximately around 5°. It will be shown in the next section that this is sufficient to be detected in an SP-STM junction. In subfigure 5.15***(b)*** the phase difference between the excitation current and the oscillation of the magnetization is depicted. It can be seen that the data follows an arc tangent behavior for all current magnitudes. This implies a phase jump of π at the resonance frequency.

The results of these calculations show that the system can be understood as a driven harmonic oscillator, similar to classical mechanics [119]. It can be seen that the phase lag is $\pi/2$ at the resonance frequency. Thus, no heterodyne resonance, as explained in chapter 3, can be detected at this frequency. But fortunately, a finite linewidth is present due to damping which will yield a signal below and above the resonance frequency f_0.

Complete signal amplitude and shape retrieved from micromagnetic simulations

With the shown simulation data it is possible to calculate the full spectra of a magnetic resonance experiment at the center of an Fe island. The signal actually detected by a Lock-in assisted magnetic resonance measurement was derived in equation 3.13 and should be recalled at this point:

$$I_{\text{Lock-in}} = \frac{1}{2} U_{\text{rf}} \frac{\partial I(U_0)}{\partial U} P_{\text{T}} P_{\text{S}} A \cos(\varphi) + \frac{1}{2} U_{\text{rf}}^2 \frac{\partial^2 I(U_0)}{\partial U^2} \tag{5.2}$$

It should be noted that the time dependence was already removed as the frequencies of the current and of the magnetic oscillation are equal. All remaining parameters can be approximated as follows: The amplitude A and the phase φ can be obtained by fitting the simulated amplitude (fig. 5.15***(a)***) with a Lorentzian and the phase difference (fig. 5.15***(b)***) with an arc tangents function. Values of the spin polarization $P_{\text{T}} P_{\text{S}}$ were obtain by spin-polarized measurements shown in section 5.2.3 ($P \approx 0.31$). The conductance $\partial I(U_0)/\partial U$ can be approximated by reasonable values of the tunneling gap. For example, tunneling at $I_0 = 300\,\text{nA}$

and $U_0 = -400\,\text{mV}$ was achieved in experiments regularly. Together with the amplitude of the rf-voltage U_{rf} this also determines the strength of the excitation current and thus, which amplitude curve (fig. 5.15*(a)*) has to be used.

The last remaining parameter is the amplitude of the rf-voltage U_{rf}. In the ideal case this parameter should be simply defined by the value set at the rf-generator. In this case the second term of the signal (eq. 5.2) is a constant offset and can be neglected. The full signal, with an actual rf-voltage amplitude of $U_{\text{rf}} = 330\,\text{mV}$, is shown in figure 5.16. A clear peak-dip pair is obtained yielding an amplitude of about $\Delta I_{\text{Lock-in}} = I_{\text{Lock-in,max}} - I_{\text{Lock-in,min}} = 280\,\text{pA}$. This is the actual magnetic resonance signal expected for the shown Fe island. It will be compared to experimentally obtained spectra in the next section.

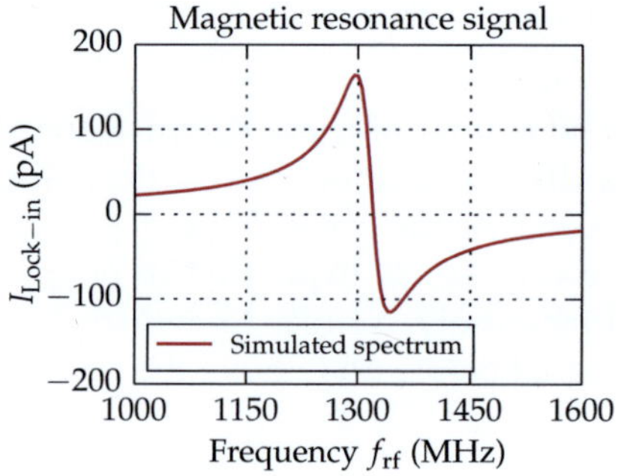

Figure 5.16: *Simulated spectra of a magnetic resonance measurement on an Fe island. Micromagentic simulations were used to obtain oscillation amplitude and phase difference at an excitation current of* $I_{\text{rf}} = 250\,\text{nA}$. *Reasonable experimental values for* P, $\partial I(U_0)/\partial U$ *and* U_{rf} *were used (given in the text). A signal strength of* $\Delta I_{\text{Lock-in}} = 280\,\text{pA}$ *is attained.*

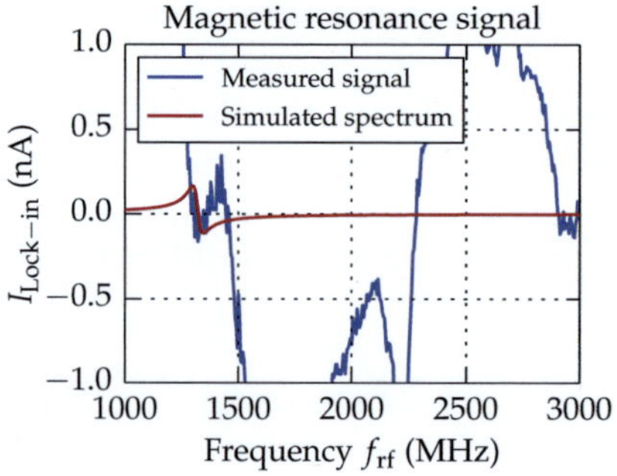

Figure 5.17: *Magnetic resonance measurement and simulation as explained in the text. The experimental curve was obtained by stabilizing the tip above a vortex core* ($I_0 = 300\,\text{nA}$, $U_0 = -400\,\text{mV}$). *The frequency at the rf-generator was swept from* 1000 *to* 3000 MHz. *The Lock-in signal acquired by amplitude modulation is plotted in blue. For comparison the simulated spectra from section 5.2.4 is overlaid.*

5.2.5 Resonance measurements

The presented simulation yields detailed information about the signal shape and amplitude. It was intended to use these values to identify the resonance in an actual measurement. However, a typical spectrum obtained on an Fe island on W(110) is shown in figure 5.17. The STM tip was fixed above the vortex core and the frequency of the rf-generator was swept. The same tunneling parameters were used for the simulation and the experiment ($I_0 = 300\,\mathrm{nA}$, $U_0 = -400\,\mathrm{mV}$). The current detected by the Lock-in amplifier (blue curve) is shown in comparison to the spectrum obtained from the simulations (red curve). It can be seen that the two spectra do not coincide at all. In the experimental spectrum many peaks and dips of huge magnitudes can be observed.

The huge difference of the experimental spectrum with respect to the simulation can be understood by reconsidering the full formula of the Lock-in signal (eq. 5.2). Especially the second term, which can not be tackled in the simulations, can be identified as the source of the vast signals: The transmission measurements in chapter 4.3 showed that the amplitude of the rf-voltage U_{rf} at the tunneling junction is not constant over frequency. This means that if a non-linear IV-curve is present ($\partial^2 I(U_0)/\partial U^2 \neq 0$), the second term in equation 5.2 will vary quadraticaly according to the transmission. This explains the huge features in the spectrum shown. Thus, the challenge is to either identify the magnetic resonance component inside the total signal or to suppress the contributions originating from transmission effects.

Signal identification due to lateral variation

The first attempt to identify the resonance signal was to probe an island at several different lateral positions. The island, along with its magnetic contrast measured by an SP-STM experiment, is shown in figure 5.18*(a)* and *(b)*. A tilted tip polarization in yz-direction is present which explains the nontrivial magnetic contrast.

Three magnetic resonance spectra, using the radio frequency method, were measured at different positions on the island (fig. 5.18*(c)*). It can be seen that the amplitude of the spectra varies depending on the position on the island. Since the amplitude of a true magnetic resonance signal should also vary laterally on an island, it can be claimed that the observed signals are resonances detected.

Unfortunately, the islands do not exhibit a laterally constant conductance (due to magnetic contrast). Thereby, also the deviation of the conductance ($\partial^2 I(U_0)/\partial U^2$) will depend on the position. Thus, the second term in equation 5.2 will yield position-dependent, not magnetic resonance related contributions to the measured signals.

To confirm this dependence on the local conductance, a full map at a peak ($f_{\mathrm{rf}} = 1550\,\mathrm{MHz}$) within the spectrum was recored (fig. 5.18*(d)*). It shows the amplitude of the signal detected by the Lock-in in an area of $80 \times 80\,\mathrm{nm}$ around the vortex core. This image can be compared to the conventional conductance map *(b)*. Great conformity is found which indicates that the peaks and the amplitude shift in the spectra in *(c)* are not caused by magnetic resonance of the island but by differences in the local density of states.

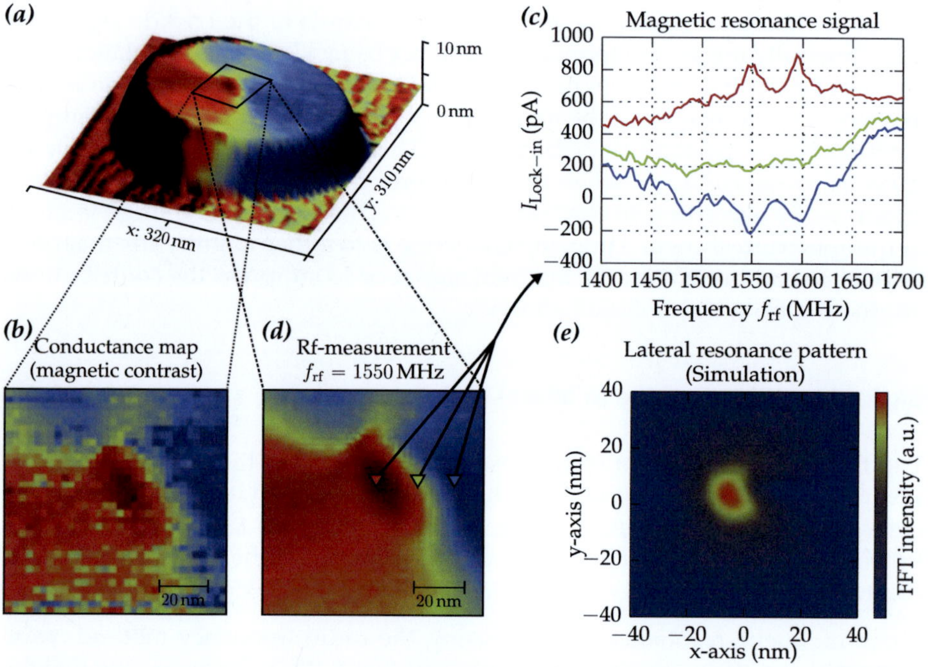

Figure 5.18: *Lateral variation of the magnetic contrast and the signal acquired by an rf-experiment.* ***(a)*** *shows the 3D representation of the island with overlaid magnetic contrast* ***(b)****. A titled tip direction is present. Three rf-experiments at different positions on the island are shown in* ***(c)*** *(I_0 = 300 nA, U_0 = −400 mV). A full map of the magnetic resonance signal at f_{rf} = 1550 MHz was acquired by fixing this frequency at the rf-generator* ***(d)****. The corresponding positions of the spectra are marked. Subfigure* ***(e)*** *shows a lateral resonance pattern obtained by simulations. The tip spin-polarization direction was fairly matched to the experimental data.*

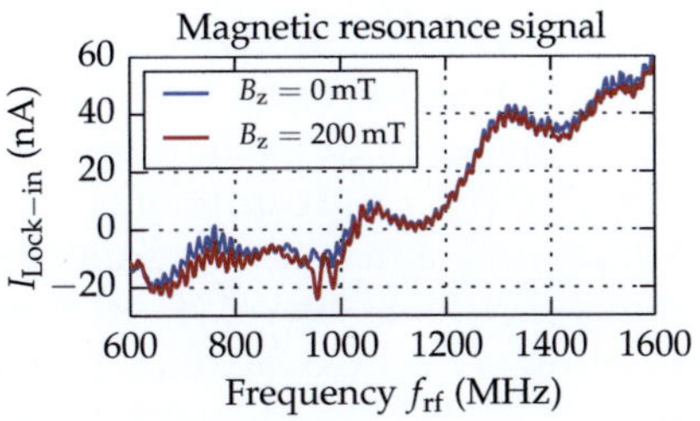

Figure 5.19: *Field dependence of the magnetic resonance data. The two curves were obtained at the same experimental conditions except for an external magnetic field in z-direction. Great care was taken to rule out tip changes between the spectra. The lateral position of the spectra on the island was matched as good as possible.*

Subfigure 5.18*(e)* shows a simulated lateral magnetic resonance pattern for a tip polarization in yz-direction. A Fourier transformation of the oscillating magnetization was performed at every point of the island surface. The intensity of these Fourier spectra at the resonance frequency is depicted. It can be seen that the pattern partially matches with the measurement (fig. 5.18*(d)*). However, it must be admitted that the global symmetry does not fit. The left and right areas of the experiment show inverted contrast, whereas the simulation predicts a symmetric amplitude around the vortex core. An explanation for this discrepancy could be the much stronger intensity of the signal contributions originating from locally different conductance. Thereby, the resonance component of signal gets superimposed and becomes insignificant in the outer areas.

It has to be admitted that through the shown detection of lateral variations only slight hints of a resonance can be found. An actual proof that a resonance was found in the island is not possible.

A potential relief could be the use of a very small bias voltage (the shown measurements were performed at $U_0 = -400\,\mathrm{mV}$, which is rather high). In the low energy range the non-linearity of the IV-curve should be negligible. Thus, only contributions originating from inelastic excitations (inelastic excitations are proportional to $\partial^2 I(U_0)/\partial U^2$) should be present. Thereby, an easier identification of the sought-after resonance peaks should be possible. But to the present day, no lateral data proofing the signal to be different from local conductance could be gained.

Signal identification by applying an external magnetic field

As explained previously, an external magnetic field in z-direction shifts the resonance frequencies of the Fe islands (ch. 5.2.4) due to Zeeman-interaction. As the experimental setup is equipped with a superconducting electromagnet it is possible to measure the islands at different magnetic fields. Hence, a shift or generally a change in the resonance spectrum is expected for different magnetic fields.

Many *in situ* experiments (same Fe island and same STM tip) with different magnetic field strengths ($0 < B_z < 200\,\mathrm{mT}$) were performed. An example of two spectra, taken at different magnetic fields on the same island, is shown in figure 5.19. First the vortex structure was confirmed by a standard SP-STM experiment. Afterwards, a magnetic resonance spectrum was taken at the position of the vortex core (blue curve). Then a magnetic field of $200\,\mathrm{mT}$ was applied. During the ramp of the field, extreme care was taken to preserve the tip state (A different tip state drastically changes the spectrum due to different density of states in the tip). This inevitably induces slight lateral drift. After repositioning the tip at the vortex core position as good as possible, another magnetic resonance spectrum was acquired (red curve). It can be seen that only minor differences are visible, e.g. at $950\,\mathrm{MHz}$. It cannot be strictly excluded that these differences are due to the shift of magnetic resonances. But a clear prove is also not possible since the repositioning of the tip might not be exact enough. Thereby, again locally different density of states of the island are encountered which distort the signal.

Transmission normalized measurements

It was shown that neither lateral variation nor the use of an external magnetic field could clearly confirm the magnetic resonance origin of (parts of) the signal. The relief is the use of the rf-amplitude normalization technique which was already explained in chapter 4.3.2: Initially, transmission data has to be recorded with the STM tip and the sample that will be studied. This data yields the required rf-amplitude to be produced at the rf-generator to achieve a certain, constant rf-amplitude at the tunneling junction. The data can be used during frequency sweeps to scale the set amplitude at the rf-generator. This will remove the frequency dependence of U_{rf}, and thus the second term of the current (eq. 5.2) will be constant. This constant offset can be removed from the data easily. All peaks or dips above the noise limit are then possible signs of resonances. Of course, lateral variation will still be present. Thus, a combined study using amplitude normalization and frequency shift by an external magnetic field is necessary.

Unfortunately, transmission normalized measurements could only be conducted at zero field within the period of this PhD work. Such an experiment is presented in figure 5.20. In subfigure 5.20***(a)*** a usual conductance map, unveiling the magnetic vortex structure, is shown. This map can be compared to subfigure 5.20***(b)***, which shows an rf-measurement at $1040\,\mathrm{MHz}$. The images were recorded at the same location. It can be seen that in the center of the vortex core an increased resonance signal is present (red color-code). The lateral shape of the signal shows a dot-like symmetry which fits to simulations (fig. 5.18***(e)***). At the edge of the image, the contrast is most likely again dominated by contributions arising from different local densities of states. This can be stated as the contrast at the outside agrees with the symmetry of the conductance map. It has to be admitted that lateral resolved resonance maps like figure 5.20*(b)* could be found in a wide frequency range. This is in contradiction to the expectations of a magnetic resonance measurement and cannot be explained by the theoretical background presented.

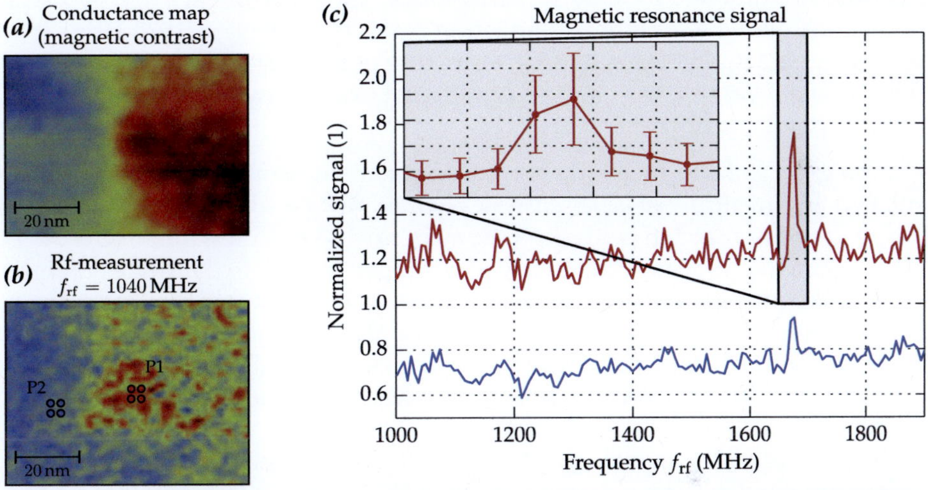

Figure 5.20: *Transmission normalized magnetic resonance measurement of Fe on W(110). Subfigure* ***(a)*** *shows the magnetic contrast unveiling the vortex structure. In* ***(b)*** *the map of rf-measurement is depicted. The increased intensity (red color-code) exhibits a dot-like shape. At positions P1 and P2 magnetic resonance spectra were recorded. They were normalized by a spectrum next to the island (not shown) and are shown in subfigure* ***(c)****. The spectrum at the vortex core (red) shows an increased intensity at* 1676 MHz*, which is a potential resonance peak (see discussion in the text). The inset shows a magnified view of the peak including the standard deviation of one sigma.*

Transmission normalized spectra were taken at the positions P1 (on the vortex core) and P2 (next to the vortex core) (fig. 5.20***(b)***). To avoid local disturbance, a set of 4 spectra was taken at both positions and each set was averaged. Additionally, a reference spectrum was taken next to the island (not shown). Due to imperfect normalization of the transmission, some effects depending on the latter are still visible in the raw data. Thus, both spectra on the island were divided by the reference spectrum next to the island. The result is plotted in figure 5.20***(c)***. Rather flat curves are obtained. The curve outside the core (blue) exhibits a constant offset from unity (from the reference spectrum) which can be explained by different local densities of states. The spectrum at the vortex core (red curve) shows a significant deviation from its flat base at 1676 MHz. This spike can potentially be a magnetic resonance of the vortex. But as mentioned, this could not yet be cross-checked at different external magnetic fields. Hence, no clear proof can be presented.

Claiming the spike at 1676 MHz to be a magnetic resonance signal, more differentiated arguments must be discussed: First, it must be admitted that the signal shape does not match a peak-dip-pair as predicted by simulations (ch. 5.2.4). Then again, the resonance frequency of the signal is close to the value obtained by simulations of about 1800 MHz. A second issue is the fact that the spike is visible in all raw data of this measurement (on the island and next to the island). Only its relative height difference yields the signal in 5.20***(c)***. This suggests a problem of the transmission normalization leading to (rapid) variations of the rf-amplitude (U_{rf}) during an rf-frequency sweep. There, particular steep slops of the transmission can easily lead to tiny offsets in frequency or in amplitude. This could yield a spike as seen in figure 5.20***(c)***, especially since a larger error is present at the data points of the spike. On the contrary, the transmission normalization data was cross-check very carefully and no inhomogeneity was found at the frequency of the spike. Thus, it can be assumed that the spike renders a real difference between the vortex core and the background. As this difference is only present at 1676 MHz, the data exactly matches the properties expected for a magnetic resonance signal. But due to the lack of magnetic field dependent data, no solid proof can be presented.

5.3 Summary magnetic resonance in Fe on W(110)

In this chapter the experiments conducted on vortices in Fe islands on W(110) were shown. This system was chosen since it is well studied and is therefore quite handy in the context of STM experiments. It fulfills all requirements to proof the concept of heterodyne magnetic resonance experiments in an STM.

It was shown by experiments and simulations that the islands exhibit a magnetic vortex structure according to previous findings. Thereto, growth and spin-polarized measurements were presented. In the next step the dynamics of the islands were simulated to gain knowledge about the resonance frequencies. This simplifies the experimental realization. As the experiments showed huge signals related to transmission effects of the rf-voltage, the interpretation of the spectra turned out to be challenging. The lateral variation of the resonance could be an approach. Unfortunately, the signals originating from different local density of states outperform the magnetic resonance signals making it impossible to proof the proposed technique. Also the verification of the technique by comparing the resonance spectra at different magnetic fields turned out to be difficult and not convincing.

A promising approach to avoid these problems is the method of normalizing the rf-amplitude at the tunneling junction. This suppresses almost all transmission effects so that only the experimental noise remains as the limiting factor. First measurement using this improved method were presented and are showing promising spectra and lateral shapes.

6 Skyrmionic spin structure in iron on iridium

In the previous chapter the first test experiments of the heterodyne magnetic resonance technique were shown. Although the vortex test system fulfills all requirements of a good proof of concept sample, the experiments turned out to be difficult and indistinct. It was shown that this is mainly due to transmission effects inside the experimental setup. But additionally, the weakness of the excitation diminishes the signal strength and increases the difficulties. It was shown that a rather high current of $500\,\mathrm{nA}$ is just enough to tilt the magnetization about $5°$. This is a clear bottleneck which could be avoided by using an easier excitable system.

For this reason the second system presented is a skyrmionic spin structure in iron on iridium (111). This magnetic structure, which was found recently in this system [120], will be introduced hereafter. The important fact is that this structure can be found in mono-atomic layers. This should lower the torque needed to excite the system since not a thick island (many nm) must be driven. Thereby, lower excitation currents should be sufficient to achieve large gyrations of the magnetic moments.

Unfortunately, drawbacks compared to the vortex structure are present as well. First of all, the introduced micromagnetic simulations cannot be utilized. This is because the studied spin structure exists on the atomic length scale where quantum mechanical effects cannot be discarded. Hence, no easy approximation of the resonance frequencies and modes is possible. This hinders the search for resonances as a wide frequency range must be probed.

The second drawback is the fact that the skyrmionic system in Fe on Ir(111) was just recently discovered in 2011 [120]. There are still unknown properties and controversies about this system. Thus, it must be admitted that this system is not well established and understood like the vortices in Fe on W(110), which were used in the previous chapter. This is a clear disadvantage because the aim of this work is to verify the new experimental method presented and not to challenge the (new) physics of the magnetic skyrmion system. Thus, misunderstood behavior during the experiments can only hardly be assigned to the method or the system. Nevertheless, the primarily measurements that were conducted on this system will be shown on this chapter.

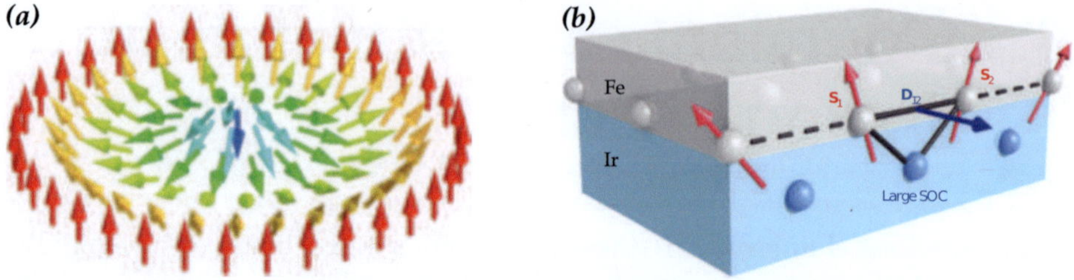

Figure 6.1: ***(a)*** *magnetic skyrmion structure. The individual arrows represent the local magnetization. At the edges they point upwards, whereas they are oriented downwards in the center. Subfigure* ***(b)*** *shows the interface induced Dzyaloshinsky-Moriya interaction (DMI) of the system Fe on Ir(111). Two neighboring spins S_1 and S_2 indirectly interact via the Ir substrate. The latter features a large spin-orbit coupling (SOC). Thus, the DMI-vector D_{12} lies in the sample plane.* ***(a)****: Figure taken from [127].* ***(b)****: Adapted by permission from Macmillan Publishers Ltd: Nature Nanotechnology [128], copyright (2013).*

6.1 Introduction of the magnetic skyrmion structure

Skyrmions are named after T.H.R. Skyrme who, used the concept of *topological solitons* to describe baryons as defects in a non-linear field theory [121–124]. Solitons are wave packages in a dispersive non-linear medium. They maintain their shape because the effects of dispersion and non-linearity compensate each other. Such wave forms were observed in water inside narrow channels already 200 years ago. The subclasses of topological solitons possess a charge-like property, and thus they cannot be annihilated or created arbitrarily. Skyrme showed that this originates from an attractive or repulsive interaction between the solitons. Hence, these solutions are called skyrmions and their charge-like property is called skyrmion number. This fundamental concept was found to be present in various other physical research areas, magnetism amongst others [125, 126].

A skyrmion structure within the scope of magnetism is shown in figure 6.1*(a)*. It can be seen that the magnetization of the shown structure rotates 360° when moving along the diameter of the magnet. This chiral spin structure cannot be continuously changed to its ferromagnetic state [128]. This renders the topological charge of this object. The reason why magnetic moments will align in such an arrangement is the competition between Heisenberg exchange and Dzyaloshinsky-Moriya interaction (DMI) [29, 30]. They were already briefly introduced in the chapter about additional terms of the spin Hamiltonian in chapter 2.1.1:

$$\mathcal{H}_{\text{ex}} + \mathcal{H}_{\text{dmi}} = J_{12}\mathbf{S}_1 \cdot \mathbf{S}_2 + \mathbf{D}_{12} \cdot (\mathbf{S}_1 \times \mathbf{S}_2)\,. \tag{6.1}$$

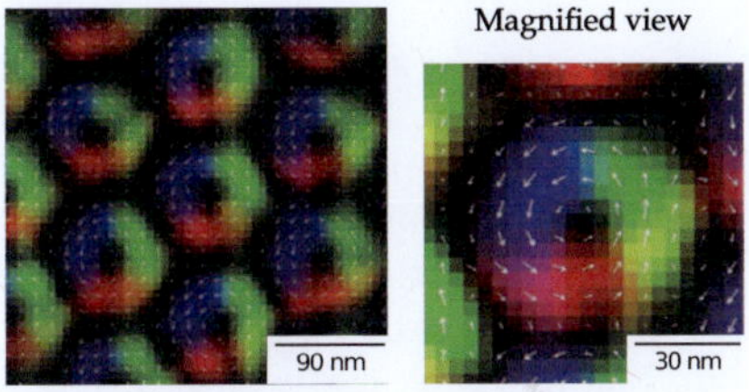

Figure 6.2: *Skyrmion lattice in FeCoSi measured by Lorentz microscopy. The system was brought to this state by an external magnetic field of* 50 mT. *A periodicity of* 90 nm *was found. The overlaid arrows indicate the direction of the local magnetization. Reprinted by permission from Macmillan Publishers Ltd: Nature [130], copyright (2010).*

$\mathbf{S}_1$ and $\mathbf{S}_2$ are the spins of two neighboring atoms, J_{12} is the Heisenberg exchange constant and $\mathbf{D}_{12}$ is the DMI-vector. The first term will favor a parallel or antiparallel alignment of the spins, depending on the sign of J_{12}. This yields ferro- or antiferromagnetism. In a system where space inversion symmetry is broken (e.g. at surfaces or interfaces) the DMI has to be taken into account (second term of equation 6.1). This interaction favors a 90° misalignment between the spins, which can be understood as they are concatenated by a cross-product. In systems where $|\mathbf{D}_{12}|$ and J_{12} have similar magnitudes, the competition between these two interactions will lead to the stabilization of complex magnetic orders like the skyrmion structure. In the system under study this is given by the broken symmetry at the interface between the Fe mono-layer and the Ir substrate. The strong spin-orbit coupling (SOC) of Ir induces a large $|\mathbf{D}_{12}|$. This yields the sketched (fig. 6.1*(b)*) three-site indirect exchange mechanism [129] which conforms to equ. 6.1.

6.1.1 Previous experimental observations

The magnetic skyrmion structure was first observed experimentally by neutron scattering in bulk MnSi, in 2009 [131]. To observe skyrmions a small external magnetic field had to be applied. This is necessary to reach the skyrmion phase from the helical ground state of MnSi. Shortly afterwards, Lorentz microscopy was used to directly image the local magnetization of a skyrmion structure in a similar bulk system [130]. Again an external field has been used to induce the skyrmions. A hexagonal lattice with a periodicity of 90 nm (fig. 6.2) was found.

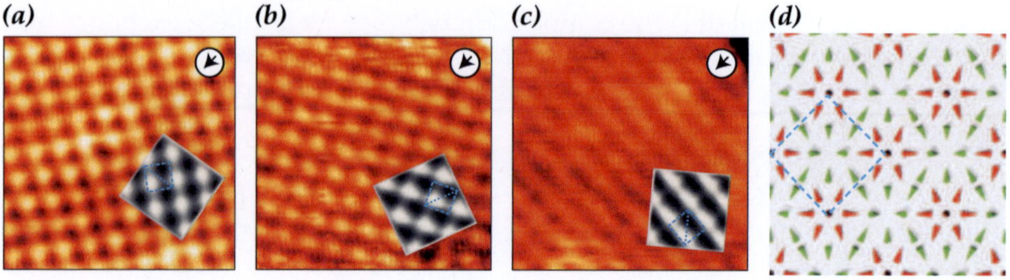

Figure 6.3: *(a)-(c) magnetic contrast of three different rotational domains recorded with the same spin-polarized tip. The tip polarization is indicated by the arrow in the upper right. All three domains show different contrasts (**(a)** and **(b)** square-like structures; **(c)** stripes). The insets show simulations based on the Tersoff-Hamann model, which confirms the measurements. **(d)** shows the nano-skyrmion structure, which can explain the observed contrast. The blue rectangle in all four images shows the magnetic unit-cell. Reprinted by permission from Macmillan Publishers Ltd: Nature Physics [120], copyright (2011).*

Already in 2006, it had been proposed that the skyrmion structure could also be the spontaneous ground state of a magnetic system [132]. This was mostly confirmed in 2011 by Heinze *et al.* using SP-STM to study the system Fe on Ir(111) [120]. This system was studied intensively before but insufficient, spacial averaging, experimental techniques were applied [133, 134] or wrong interpretations were made [135, 136]. In the study of Heinze *et al.* an in-plane spin-polarized Fe tip was used. They found three different magnetic superstructures on the first atomic layer of Fe.

These three domains are shown in figure 6.3*(a)-(c)*. It must be emphasized that the spin polarization of the tip was retained for all three pictures, which is indicated by the small arrows in the upper right. The three domains are caused by different rotations (around 120°) of the magnetic structure with respect to the Ir substrate. Thereby, the fixed spin polarization of the tip yields different magnetic contrasts for each domain (the contrast can be understood as a projection of the local magnetization onto the tip polarization). Simulated images of the magnetic contrast are overlaid. They fit the observed square like ***(a)***, ***(b)*** and stripe like ***(c)*** superstructures.

By appropriated rotation and suitable color-coding of the obtained data a three-dimensional model of the magnetism could be constructed [137]. This model is shown in figure 6.3*(d)*. The local magnetization is depicted by the cones, whereas the color indicates the z-component. These are so-called *nano-skyrmions* with a lattice periodicity of only about 1 nm. This model explains the magnetic contrast observed of all three rotational domains (fig. 6.3*(a)*-*(c)*). The small

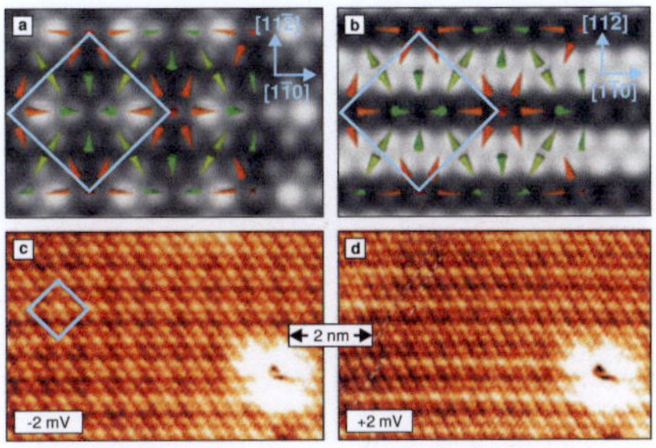

Figure 6.4: *Simulations ((**a**) and (**b**)) and atomic resolution measurements ((**c**) and (**d**)) showing the TAMR on Fe on Ir(111) measured with a non-magnetic tip. At the bias voltage of* $-2\,\mathrm{mV}$ *a square-like superstructure is obtained ((**a**) and (**c**)), whereas a bias voltage of* $+2\,\mathrm{mV}$ *yields stripes ((**b**) and (**d**)). The blue rectangle depicts the magnetic unit-cell. Reprinted figure with permission from [138]. Copyright (2012) by the American Physical Society.*

lattice size is caused by the large DMI at the interface between Fe and Ir. The blue rectangle in all four images of figure 6.3 shows the magnetic unit-cell of the structure.

In this system an additional contrast was found while scanning at small bias voltages [137, 138]. This contrast, which can also be observed using a non-spin-polarized tip, originates from an effect called tunneling anisotropic magnetoresistance (TAMR) [139]. Figure 6.4 shows simulations ((*a*) and (*b*)) and corresponding STM measurements ((*c*) and (*d*)). The measurements show atomic resolution and an additional superstructure. This effect is bias-dependent and can be explained by strong spin-orbit coupling which modulates the local density of states. For instance, in figure 6.4(*b*) and (*d*) the situation at low positive bias voltage (+2 mV) is shown. Stripes along the $[1\bar{1}0]$ direction with an increased conductance were observed. This can be explained in the case an alignment of the magnetic moment along the crystallographic $[1\bar{1}0]$ direction generates a lower electronic density of states and an alignment perpendicular to this axis features an increased one. Hence, a spin-related effect is observed despite a non-spin-polarized tip is used. The figures show the difference between two bias voltages. In (*a*) and (*c*) a square-like pattern can be observed, whereas in (*b*) and (*d*) the stripes mentioned can be seen. The blue rectangle again shows the magnetic unit-cell.

In the STM images in figure 6.4 atomic resolution could be achieved additionally (small periodicity in (*c*) and (*d*)). This allows to gain new insights into the local magnetism on the surface: By carefully comparing the periodicity of the experimental atomic resolution and the contrast of TAMR to simulations of the skyrmion lattice, it was shown that the atomic and the magnetic structure are incommensurable [120,137]. This means that the magnetic structure in 6.4(*a*) and (*b*), where each cone represents one Fe atom, is not valid but has to be slightly modified. The magnetic lattice is about 10 % compressed compared to the model where one atom caries one cone in figure 6.3. It should be noted that this proof was only possible by comparison to simulations since no measurements where all three effects (atomic-, magnetic- and TAMR-resolution) could be observed simultaneously were published yet.

The shown *nano-skyrmion* structure in Fe on Ir(111) is the target of measurements of magnetic resonance. To the knowledge of the author, no dynamic measurements of this system were conducted yet. Therefore, the dynamics of skyrmions will be reviewed on a more general scope in the following chapter.

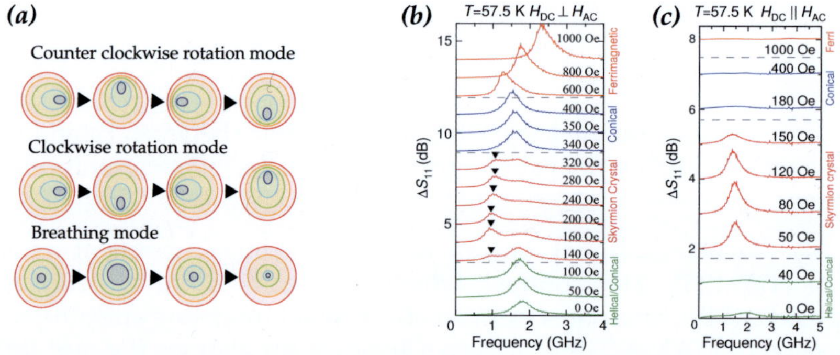

Figure 6.5: ***(a)*** *Illustration of three different dynamical modes of magnetic skyrmions. These modes were predicted by numerical simulations. The color contours represent the z-component of the magnetization.* ***(b)*** *and* ***(c)*** *show microwave response experiments of a* Cu_2OSeO_3 *crystal.* ***(b)*** *in the perpendicular (rf-field in the plane) and* ***(c)*** *in the parallel configuration (rf-field out of the plane). The red color denotes the skyrmion phase. In* ***(b)*** *the rotational mode is marked. In* ***(c)*** *the breathing mode is observed as a clear peak. Reprinted figures with permission from [146]. Copyright (2012) by the American Physical Society.*

6.1.2 Dynamics of magnetic skyrmions

Following various studies of current-induced dynamics of a skyrmion lattice [140–144], Mochizuki proposed three different spin-wave modes inside a skyrmion crystal [145], in 2012. He used a classical Heisenberg model to account for Exchange-, Dzyaloshinskii-Moriya- and Zeeman-interaction. Thereby, the LLG equation was solved numerically on a grid of 288×288 sites. He achieved two rotational modes (clock- and counterclockwise) and a breathing mode. Their frequencies all lie in the low GHz range. Figure 6.5*(a)* depicts these three modes schematically. In the two rotational modes the core of the skyrmion rotates around its equilibrium position (the center). These modes are similar to the gyroscopic mode of the vortex structure (ch. 5.1.2). The breathing mode consists of periodic expansion and shrinkage of the skyrmion core.

These modes could be experimentally verified in the same year in Cu_2OSeO_3 [146]. This multiferroic material features a magnetic skyrmion lattice with a periodicity of about 50 nm [147]. A microwave was irradiated onto the single crystal and the absorption was measured by a vector network analyzer. To account for transmission effects, the background acquired at high field was subtracted. The spectra for different external magnetic fields are shown in figure 6.5*(b)* and *(c)*. In subfigure *(b)* the spectra for an exciting field perpendicular to the external field are depicted (in-plane rf-field). The resonance peak at about 1 GHz in the skyrmionic phase (red) could be attributed to the counterclockwise rotation mode. The clockwise mode could not be clearly identified. In *(c)* the spectra for a parallel alignment between static and exciting field are shown (rf-field out-of-plane). They clearly unveil the resonance peak of the breathing mode at about 1.5 GHz.

The shown dynamic modes are the template and motivation to search for such modes by means of the magnetic resonance technique presented earlier in this work (ch. 3). The reviewed findings feature resonance frequencies in the accessible range of the present experimental setup. Unfortunately, there are, to the knowledge of the author, no publications related to the dynamics of the *nano-skyrmion* system Fe on Ir(111) yet. This derogates its use as a proof of concept sample since there is no data on actual resonance frequencies. Nevertheless, the prospect of strong excitation of the thin Fe layer validates preliminary test experiments: Even small current densities allow to move or rotate skyrmion lattices in bulk (the critical current densities are 5 orders of magnitude smaller for moving a skyrmion than moving a domain wall) [140–144]. This corroborates the assumption that in thin films, like the Fe on Ir system, even lower currents are sufficient to excite a magnetic rotation or breathing mode via the spin-torque effect.

6.2 Experiments

6.2.1 Preparation and Growth

To measure magnetic resonance in skyrmions in Fe on Ir(111), it is necessary to initially reproduce this system. Therefore, a 99.99 % pure Ir single crystal was purchased from MaTecK GmbH. The manufacturer states that the main impurity is C, with about 3 ppm.

To clean this crystal the methods decribed in [89, 148] were used: Repeated cycles of Ar^+ sputtering at 3 kV, annealing up to 1500 °C and heating in O_2 atmosphere were carried out. These treatments remove contaminations from the bulk of the crystal. The final procedure, before insertion of the sample into the STM, consisted of only 20 min of sputtering and a 30 s high temperature flash at 1500 °C. This removes remaining adsorbates, which often show up as superstructures (not shown), and yields a clean crystal surface (fig. 6.6*(a)*).

After this cleaning procedure thin Fe films were grown by molecular beam epitaxy (ch. 4.1.3). About two thirds of a mono-layer were deposited as soon as the sample was cooled down to about 130 °C. An STM topography image is shown in figure 6.6*(b)*. It can be seen that first and second mono-layer islands were obtained. In addition to island-growth the Fe atoms diffuse to the step edges of the Ir crystal and form elongated stripes. Hence, buried step edges of the substrate can be observed as they exhibit a height difference of about 65 pm at standard tunneling parameters ($U_0 = 1\,\mathrm{V}$ and $I_0 = 1\,\mathrm{nA}$). Furthermore, some islands show a triangular shape which indicates pseudomorphic growth following the Ir substrate. This agrees with reference [133] where a pseudomorphic growth up to the fourth mono-layer and a following Kurdjumov-Sachs transition to bulk Fe body-center-cubic growth was observed.

The presented sample structure is a perfect starting point for the confirmation of the magnetic skyrmion structure according to [120]. This subsequent step is indispensable for the approach towards magnetic resonance measurements in skyrmions.

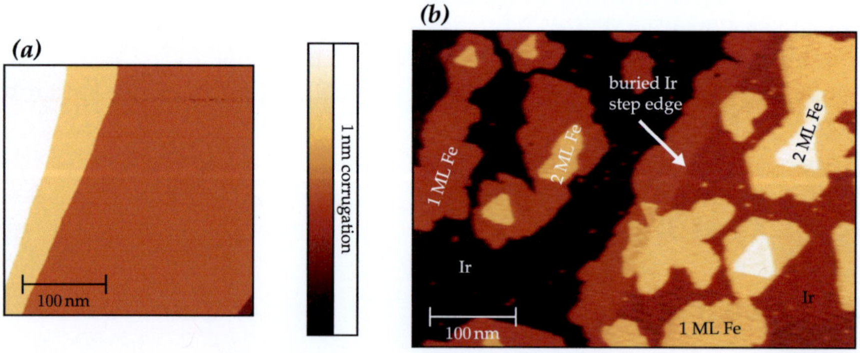

Figure 6.6: *Preparation and growth of Fe on Ir(111) measured with a non-magnetic tip. In **(a)** the clean Ir single crystal is shown. Wide terraces separated by atomic steps can be observed. Subfigure **(b)** shows the epitaxial growth of Fe on the clean Ir crystal. Fe islands up to a thickness of two atomic layers can be seen. Additionally, Fe-atoms accumulate at the buried Ir step edges. Tunneling parameters of both images:* $U_0 = 1\,\mathrm{V}$ *and* $I_0 = 1\,\mathrm{nA}$

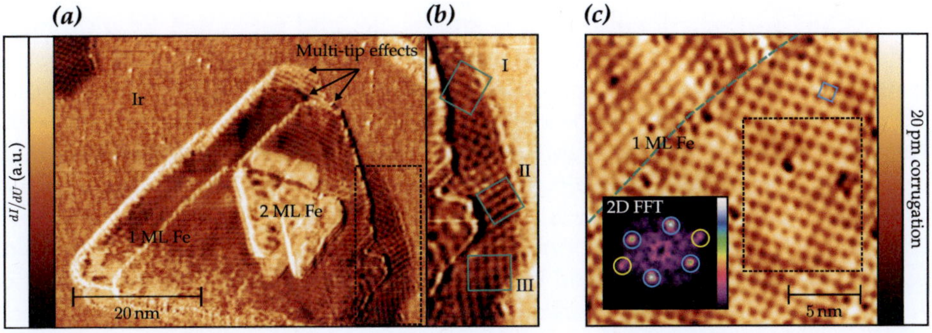

*Figure 6.7: Spin-polarized STM measurements of the system Fe on Ir(111), utilizing an Fe coated W tip. **(a)** One and two mono-layer thick triangular islands are visible. A multiple tip is present. On the first mono-layer different domains of a magnetic superstructure can be observed. **(b)** Magnified cutout of **(a)**. Three domains (I, II and III) rotated by* $120°$ *are marked by the green rectangles. Tunneling parameters in **(a)** and **(b)**:* $U_0 = 20\,\mathrm{mV}$, $I_0 = 10\,\mathrm{nA}$. *Subfigure **(c)** shows the magnetic contrast of two domains separated by the green line. The magnetic unit-cell is marked in light-blue. A 2D FFT was applied to the area inside the dashed box. The spots of the magnetic unit-cell are marked in light-blue, and the ones of the TAMR contrast are marked in yellow. Tunneling parameters in **(c)**:* $U_0 = 20\,\mathrm{mV}$, $I_0 = 1\,\mathrm{nA}$

6.2.2 Magnetism - spin-polarized measurements

After cleaning the Ir crystal and the epitaxial growth of single atomic Fe layers a spin-polarized tip was prepared. Again an Fe coated W tip was chosen as high spin polarizations could be produced reliably. The preparation of such a tip was described in chapter 4.1.3. Such a tip will most likely yield an in-plane magnetic contrast.

Figure 6.7*(a)* shows a measurement of a triangular Fe island. A map of the differential conductance is shown as it yielded a better contrast in this measurement. Unfortunately, a multi-tip is present which pictures every feature of the surface 3-4 times. At least two to three different superstructures on the first Fe layer are visible. They originate from different magnetic domains similar to the model system explained in the previous section (ch. 6.1.1). In subfigure 6.7*(b)* the magnified cutout, marked by the dashed box in *(a)*, is shown. In this area no multi-tip effects are present. All three different domains (I - III) are marked by the green rectangles. They are rotated by exactly 120° with respect to each other. It should be noticed that the rotation of all three domains fits exactly to the alignment of the rectangles. Therefore, the magnetic *nano-skyrmion* structure presented in the previous section could be confirmed.

A detailed image of higher quality is presented in 6.7*(c)*. Since an atomically flat surface is present, the magnetic contrast can be visualized by the topography data. Only two rotational domains are visible, and separated by a domain boundary depicted in green. The superstructure in the upper left domain shows a stripe-like pattern, whereas the lower right one shows a square lattice. Both superstructures exhibit a periodicity of $a \approx 1.06\,\mathrm{nm}$ which is agreeing well with reference [137]. The magnetic unit-cell is marked in light-blue. Inside the dashed box in figure 6.7*(c)* a 2D Fast-Fourier-Transformation (FFT) was performed for deeper analysis. The outcome is shown as an inset. Four strong spots, marked in blue, appear which correspond to the magnetic unit-cell. Additionally, two spots, rotated by $45°$ and scaled by a factor of $\sqrt{2}$ with respect to the magnetic unit-cell, show up and are marked in yellow. They belong to the TAMR effect which is marginally visible in the real-space image. This result also accords with the findings explained in section 6.1.1.

Therefore, the *nano-skyrmion* structure is well reproduced. This is the ideal starting point for magnetic resonance experiments as the intended magnetic structure of the sample was verified and also the spin polarization of the tip was proven.

6.2.3 Magnetic resonance measurements

The first magnetic resonance experiments were performed by stabilizing the tip above the skyrmion structure and sweeping the rf-voltage. The current of the Lock-in, which comprises the magnetic resonance signal as explained in chapter 3, was recorded. The same problem, as explained in the previous chapter of magnetic vortices (ch. 5.2.5), naturally occurred: The non-constant transmission along with the non-linearity of the IV-curve adds huge contributions, proportional to the local density of states. These contributions outperform the signals originating from magnetic resonance.

An example is shown in figure 6.8*(a)* in the upper panel. The red and the blue curves were measured at different locations P1 and P2. It can be seen that the overall shape of the curves does not vary significantly amongst the two points. In the lower panel the difference between the two curves is depicted. Only at the huge peaks between 1200 and $1500\,\mathrm{MHz}$ a difference of some hundred pA is present. A map of the signal at approximately $1200\,\mathrm{MHz}$ is shown in subfigure 6.8*(c)*. Since a lateral variation of the magnetic resonance signals can be expected, it could be claimed that these signals originate from magnetic resonances of local modes of the skyrmion lattice. But it has to be admitted that this image reproduces the magnetic contrast obtained by topography data (fig. 6.8*(b)*) quite well. Hence, the magnetic resonance in the skyrmions cannot be easily confirmed and it has to be acknowledged that the observed peaks *(a)* are most likely dominated by transmission effects. Thus, the image in 6.8*(c)* shows a conductance map measured through rf-modulation of the bias voltage.

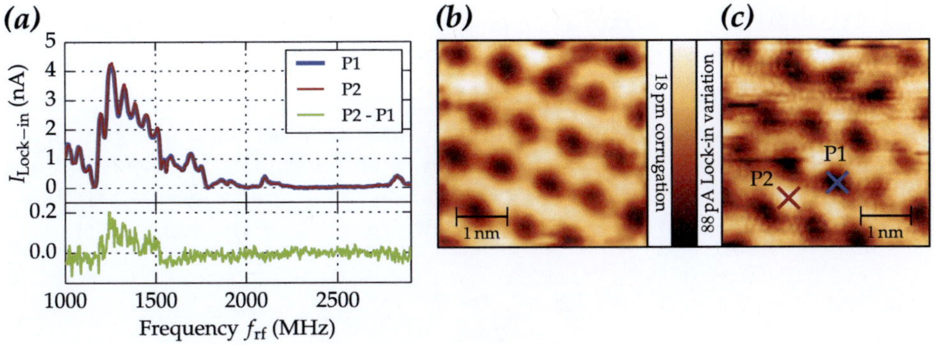

***Figure 6.8:** rf-based magnetic resonance measurement of Fe on Ir(111) using an Fe coated W tip. **(a)**: Two spectra at different lateral positions are shown. In the lower panel the difference of the two curves is depicted. Tunneling parameters: $I_0 = 20\,\mathrm{nA}$, $U_0 = 50\,\mathrm{mV}$ and $U_{\mathrm{rf}} \approx 500\,\mathrm{mV}$. Subfigure **(b)** shows the topography data yielding magnetic contrast in the scanned area. Figure **(c)** depicts the signal obtained through an rf-based measurement in the same area as **(b)** ($f_{\mathrm{rf}} = 1200\,\mathrm{MHz}$). The positions of the spectra of **(a)** are marked.*

Simultaneous observation of three signal contributions

The possibility to obtain, e.g. magnetic information of the sample by rf-modulation, was shown above. Rf-measurements at low bias voltage ($U_0 = 10\,\mathrm{mV}$) and an excellent tip allowed to extract even more information out of a single map recorded: Therefore, again 2D-FFT was utilized to examine the scan shown in figure 6.9***(a)***. The real-space image shows a complex structure which can be decomposed into several sub-lattices. For this, the corresponding areas of the Fourier transformed image were segregated and transformed back to real space (fig. 6.9***(b-d)***).

Three different contributions could be extracted. Subfigure 6.9***(b)*** shows the hexagonal atomic lattice. These are the Fe atoms following the Ir(111) surface lattice. The inset depicts the selected part of the Fourier transformed data, which in this case is also a hexagon. The obtained atomic nearest neighbor distance (NN: 274 pm) yields a lattice constant of about 383 pm, which is in good agreement with literature values [149]. The second contrast that could be extracted is the magnetic one originating from spin-polarized tunneling depicted in 6.9***(c)***. This contrast, which is the most prominent signal in the raw data, shows a square lattice. The magnetic unit-cell of about 1 nm coincides with previously shown measurements. In the third subfigure 6.9***(d)*** a stripe pattern which corresponds to TAMR can be seen. It could also be extracted from the raw data via FFT. The periodicity of the pattern of about 725 pm matches the magnetic unit-cell (fig. 6.9*(c)*) as demanded by a factor of $\sqrt{2}$. Additionally, the rotation of $45°$ between *(c)* and *(d)* evidence the extracted patterns.

The presented ability of the rf-based measurements to detect different contrast simultaneously, shows the high sensitivity obtained in this kind of experiments. This is fine, although the signals are obviously still dominated by contributions originating from non-magnetic resonance terms. Thus, measurements utilizing the previously shown transmission normalization will be presented next.

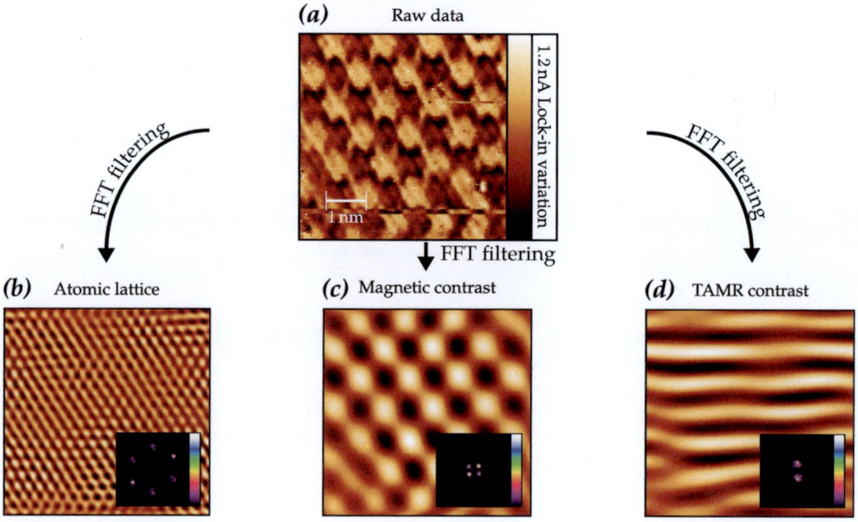

Figure 6.9: *Simultaneous measurement of atomic, magnetic and TAMR information of Fe on Ir(111). Subfigure **(a)** shows the raw data obtained through an rf-based STM measurement utilizing an Fe coated W tip. Tunneling parameters:* $I_0 = 10\,\text{nA}$, $U_0 = 10\,\text{mV}$, $U_{\text{rf}} \approx 7\,\text{mV}$ *and* $f_{\text{rf}} = 100\,\text{MHz}$. *Appropriated filtering of the Fourier transformed data yields the subfigures **(b)-(d)**. They show the real-space images and the Fourier transformed data as an inset. The atomic contrast **(b)** exhibits a hexagonal lattice with a next neighbor distance of* 274 pm. ***(c)** depicts the magnetic contrast by spin-polarized tunneling. The lattice constant of this quadratic structure is about* 1 nm. *Subfigure **(d)** shows the stripe-like pattern originating from TAMR contrast. A periodicity of* 725 pm *was observed.*

Magnetic resonance after transmission normalization

To extract the magnetic resonance based contributions from the signal, the transmission normalization as explained previously (ch. 4.3.2 and 5.2.5) was utilized. Since already minor changes of the electronic or magnetic state of the tip yield significant changes of the transmission, great care was taken to avoid any modification of the tip during and after the calibration.

A topography image, yielding acceptable magnetic contrast, is shown in figure 6.10*(a)*. Normalized magnetic resonance spectra were obtained at the three different locations marked by the colored dots. The correspondingly colored spectra are depicted in subfigure 6.10*(b)*. It can be seen that the curves are shifted which originates from different density of states at the different positions. Furthermore, the main features (e.g. the step at 1200 MHz) coincide in all three curves and are most likely caused by imperfect transmission normalization. Thus, for reasonable comparison of the curves another normalization step is necessary. Therefore, the curve colored in purple was supposed to be the background. All curves were divided by the background to uncover tiny differences. The result is shown in figure 6.10*(c)*, where the purple curve is naturally 1. Both remaining curves are still shifted, but additionally gentle oscillations are visible. The green and the blue curve show the same trend which can be explained by the close vicinity of the lateral positions. It cannot be excluded that the structure observed originates from quite broad ($\approx$ 500 MHz) resonances of the skyrmion lattice. Since no theoretical, simulated reference is available, it is hard to identify resonances. Unfortunately, such data could not be obtained for different magnetic fields to proof the magnetic origin of the signal. Only by shifting the observed structure through a magnetic field, a clear proof would be found. But without the field dependence still other sources of the signals can be found, e.g. imperfection of the transmission normalization.

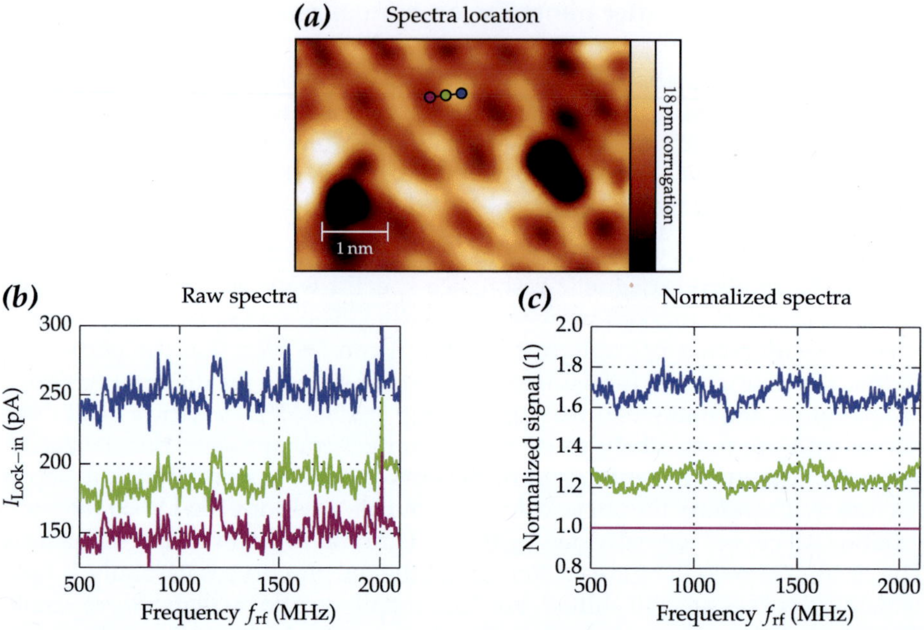

Figure 6.10: *Transmission normalized magnetic resonance of Fe on Ir(111) utilizing an Fe coated W tip. In* ***(a)*** *the area of the measurement is shown by a topography image yielding magnetic contrast. The locations of the subsequent spectra are marked by colored circles. Three rf-based magnetic resonance spectra are shown in* ***(b)****. Tunneling parameters: $I_0 = 5\,\text{nA}$, $U_0 = 5\,\text{mV}$, $U_{\text{rf}} \approx 5\,\text{mV}$. The amplitude of the radio frequency voltage was adjusted to maintain its values at the tunneling junction by previously acquired transmission data. Part* ***(c)*** *shows the normalized curves. The purple one was supposed to be the background.*

6.3 Summary magnetic resonance in Fe on Ir(111)

In the present chapter the preliminary magnetic resonance measurements of *nano-skyrmions* in Fe on Ir(111) were presented. During the introduction it should have become clear that this rather novel magnetic structure is a promising system to measure magnetic resonance by the introduced heterodyne detection scheme. On the other hand, the diversity and complexity of the system impedes quick, clear results. Especially the lack of simulation abilities reduces the usability of this novel sample for the new experimental concept of this work.

Nevertheless, some impressive measurements could be achieved. It could be shown that under excellent tunneling conditions it is possible to measure three different contrasts (atomic, magnetic and TAMR) simultaneously. This was not achieved in previous measurements. But the main attention should be given to the transmission normalized measurements. By the appropriated modification of the amplitude produced by the rf-generator, very flat and homogeneous spectra could be achieved. This gives hints to broad resonances in the skyrmion lattice which should be confirmed by further studies at different magnetic fields.

7 Resonance experiments on paramagnetic molecules

In the two previous chapters the resonance experiments on ferromagnetic samples, where exchange interactions couple the individual spins to a continuous magnetization, were presented. By contrast, the current chapter shows preliminary results of magnetic resonance in paramagnetic, organic molecules. There the magnetism is determined by the 3d electrons of an embedded metal atom. This chapter is therefore closer related to ESR.

The reason for using organic molecules is the probable decoupling of the magnetic moment from its surroundings, especially from the substrate. The organic ligands are the binding and simultaneously the shielding between the spin center and the substrate [11]. This is especially necessary since in a non-decoupled environment, conduction electrons of the substrate will continuously scatter with the molecular spin, creating a singlet ground state, i.e. $S = 0$. This is known as the Kondo effect [150–152] and should be avoided since very short spin-lifetimes are present in this case.

A second issue is the zero field splitting caused by the crystal field (ch. 2.1.1). This will easily shift the resonance frequency out of the range of the rf-generator ($f_{\mathrm{max}} = 3\,\mathrm{GHz}$). A convenient workaround is the use of spin-$1/2$ molecules. For such spin systems Kramer's theorem applies (ch. 2.1.1), which dictates the degeneracy in the absence of a magnetic field [25]. Thus, crystal field splitting can be neglected when estimating resonance frequencies. Accordingly, the molecule presented in the following section is supposed to exhibit a spin-$1/2$.

7.1 Introduction of oxyvanadium phthalocyanine

The type of molecules used in this work are the well-studied phthalocyanines (Pc) [86,152]. They are mainly used as dyes [153] and consist of four pyrrole-like lobs, connected by a ring of N and C atoms (fig. 7.1) [154,155]. In the basic phthalocyanine type, two H atoms are placed at the center of the molecule. This is called hydrogen phthalocyanine (H_2Pc), whose ball and stick model is shown in figure 7.1*(a)*.

The two central H atoms can be replaced by a metal ion providing the molecule with new electronic and magnetic properties [156]. For the purpose of magnetic resonance measurements an ion with $S = 1/2$ is preferred, as mentioned before. This can be achieved by usage of a VO^{2+}-ion as the metal center, forming oxy-vanadium phthalocyanine (VOPc). The structure of this molecule is shown in figure 7.1*(b)*. It can be seen that the VO^{2+}-ion replaced the two hydrogen atoms in the center of the metal-free phthalocyanine. The O atom protrudes from the flat structure of a normal H_2Pc.

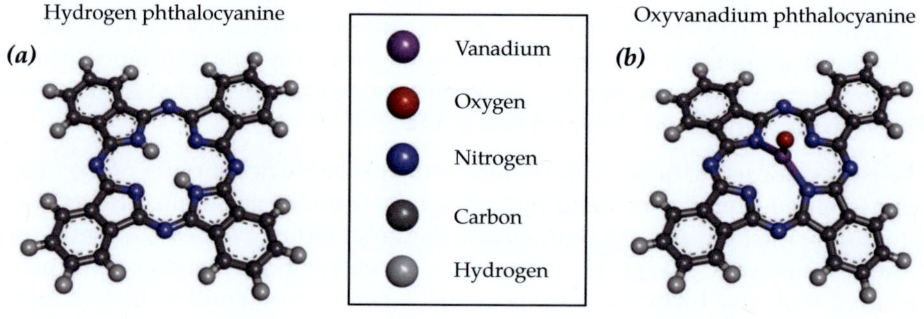

Figure 7.1: *Ball-and-stick models of H_2Pc **(a)** and VOPc **(b)**. The rather flat molecules are composed of four organic ligands attached to an aromatic ring of C and N atoms.*

The magnetism of the molecules is thereby determined by the VO^{2+}-ion. This ion consists of an O^{2-} and a V^{4+}. The neutral V atom exhibits an electronic ground state of $[Ar]3d^34s^2$. Thus, after oxidation the ion exhibits a single unpaired 3d electron. Since quenching of the orbital momentum occurs, a single electron without orbital contributions remains [157]. Conventional ESR experiments yield a g-factor of almost 2 [158]. Additionally, the V nucleus exhibits a nuclear magnetic moment of $I = 7/2$. This yields a hyperfine splitting of the resonance levels (eq. 2.16). The coupling parameters and the g-factors obtained from studies on powders of this compound [158] were used to solve the spin Hamiltonian of this system (ch. 2.1.1):

$$\mathcal{H}_{\text{eff}} = \mu_B \mathbf{B} \cdot \bar{\bar{g}} \cdot \mathbf{S} + \mathbf{I} \cdot \bar{\bar{A}} \cdot \mathbf{S}. \tag{7.1}$$

The resulting term schema is shown in figure 7.2. Since the hyperfine coupling between the electron and the nuclear spin is non-isotropic already at zero field, 9 different levels are present (7 of them are doubly degenerated). It can be seen that the external field B_0 lifts all degeneracy and splits all 16 states into two branches with $m_s = +1/2$ and $m_s = -1/2$. For the magnetic field strengths shown, the transition energies between the two branches are in the low GHz regime, as expected for a spin-1/2. These resonance frequencies are in the accessible range of the present setup, which renders the VOPc a well-suited means to test the heterodyne magnetic resonance technique.

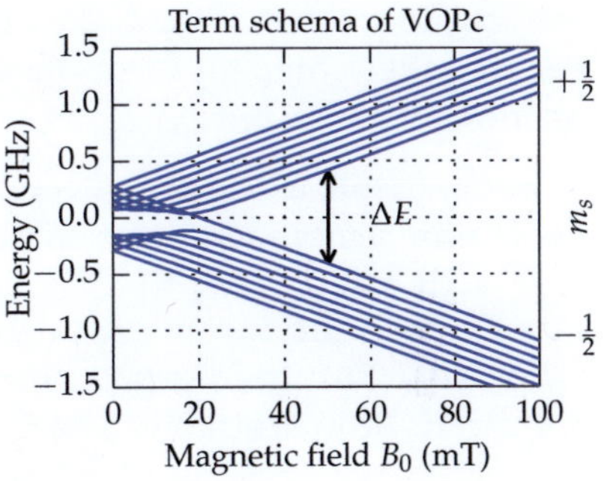

Figure 7.2: *Term schema of VOPc calculated by the Hamiltonian consisting of electron Zeeman interaction and hyperfine coupling. $S = 1/2$ and $I = 1/2$. At low fields the competition between the two terms yields a complex evolution of the states. At higher fields the Zeeman interaction dominates. Thus, two branches for the two states of the electron spin are obtained.*

The drawback of the shown molecules can be seen right away: The central ion (VO^{2+}) is only coordinated by two nitrogen atoms and the top and the bottom are exposed (fig. 7.1). This is disadvantageous as conduction electrons of the metal substrate might interact with the VO^{2+}-ion, altering or destroying its magnetic properties. A possible relief is the deposition of many molecular layers to decouple the topmost molecules from the substrate electronically [159]. To furthermore decrease the interaction amongst the molecules themselves, one approach is to utilize a mixture of nonmagnetic H_2Pc and magnetic VOPc .

Finally, it can be stated that the benefit of using phthalocyanines is their easy, robust handling under UHV conditions. They have been known to be evaporable for many decades [160] and were also studied successfully in the present experimental setup [86].

7.2 Experiments on phthalocyanine multilayers on Ag(111)

7.2.1 First molecular layer

An Ag(111) single crystal was cleaned by repeated cycles of Ar^+ sputtering and annealing to 450 °C. Thereby, an atomically clean surface, showing single atomic steps, was obtained (fig. 7.3(*a*)). As mentioned before, a decoupling by using a mixture of H_2Pc and VOPc is intended. To be able to identify the different molecule types of the mixture, both types of molecules were briefly studied before dilution of the latter.

In figure 7.3*(b)* an STM topography image of H_2Pc molecules on Ag(111) is shown. The molecules were thermally evaporated from a Knudsen-Cell. An evaporation temperature of 320 °C during a deposition time of 100 s yields a decent coverage. Single molecules and a patch of ordered molecules can be seen. The four lobes of the H_2Pc can be clearly identified. Furthermore, it can be seen that the centers of the molecules show an "empty" cavity. This is a clear signature of H_2Pc.

For comparison, a sample with pure VOPc is shown in figure 7.3*(c)*. The same evaporation temperature was chosen, but the deposition duration was extended to 150 s. It can be seen that significantly less molecules are presented. This is contrary to the lower vapor pressure expected for VOPc with respect to H_2Pc [161]. But it is practical for studies of mixtures where a small ratio of VOPc to H_2Pc is desirable. The observed shape of VOPc is identical to that of H_2Pc except at the center of the molecules. The latter shows a bright dot which corresponds to the VO^{2+}-ion. The whole VOPc structure fits the topography data very well, as can be seen in the magnified view in figure 7.3*(d)*. Due to the low coverage in figure 7.3*(c)*, three different rotations of the molecules can be observed (red arrows fig. 7.3*(c)*). They match the three-fold symmetry of the Ag(111) substrate.

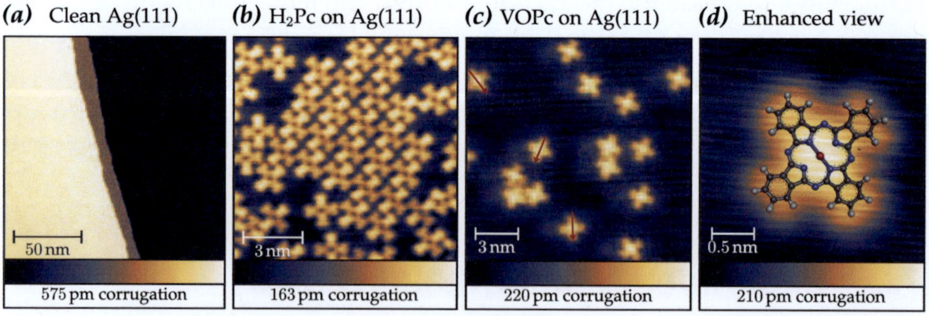

Figure 7.3: *STM topography data showing the growth of H_2Pc and VOPc on Ag(111).* ***(a)****: Atomically clean Ag crystal. Tunneling parameters: $U_0 = 1\,\mathrm{V}$ and $I_0 = 1\,\mathrm{nA}$. In subfigure* ***(b)*** *the growth and structure of H_2Pc can be seen.* ***(c)*** *and* ***(d)*** *show a sample of VOPc. The ball-and-stick model is superimposed in* ***(d)****. Tunneling parameters* ***(b)****-****(d)****: $U_0 = 300\,\mathrm{mV}$ and $I_0 = 100\,\mathrm{pA}$.*

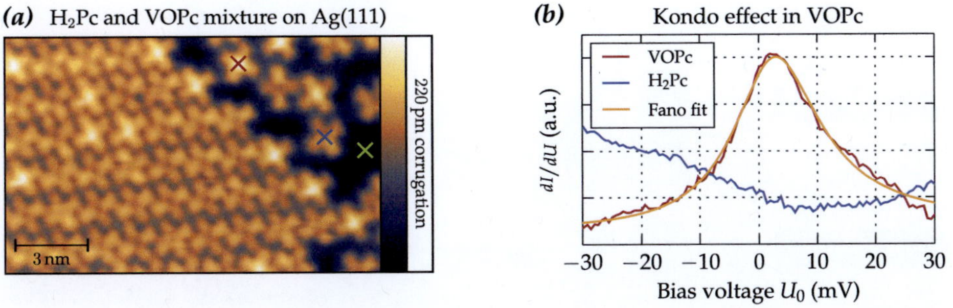

Figure 7.4: ***(a)***: *STM topography of a sample generated by mixed deposition of H_2Pc and VOPc on Ag(111). An almost fully-covered monolayer of molecules can be seen. The two different molecule types can be clearly identified as the VOPcs exhibit a bright dot in the center. Tunneling parameters:* $U_0 = 320\,\mathrm{mV}$ *and* $I_0 = 110\,\mathrm{pA}$. ***(b)***: *Tunneling spectroscopy shows a Fano lineshape, indicating a Kondo effect, only at the VOPc (The background at the green position in* ***(a)*** *was subtracted). The position of the spectra are marked in* ***(a)***.

Subsequently to the tests using the two pure molecule types, a mixture of about 50 % H_2Pc and 50 % VOPc was produced and deposited onto the previously clean Ag crystal. The mixed deposition can be observed well in figure 7.4***(a)***. Again single molecules and a closed patch of molecules can be seen. About 23 % of the molecules exhibit a bright center indicating a VOPc . The uneven ratio is perfect to find separated VOPcs.

As mentioned before, a single electron spin exposed to conduction electrons of a metal will produce a Kondo resonance [162–165]. This resonance shows up as a Fano peak at the Fermi energy in the density of states obtained by tunneling spectroscopy. Such tunneling spectra were obtained at the three positions marked by the crosses in figure 7.4***(a)***. The red cross is located at the center of a VOPc, the blue one at the position of an H_2Pc and the green one was recorded as a reference on the Ag substrate. After subtraction of the Ag background, the curves shown in figure 7.4***(b)*** were obtained. It can be seen that the VOPc, in contrast to H_2Pc, shows a strong peak at the Fermi energy. The Fano function could be fitted to the peak yielding a Kondo temperature of $T_K = 104\,\mathrm{K}$.

The observation of the Kondo resonance in VOPc is sound as it confirms the single unpaired electron of the embedded metal ion. But it also affirms the assumption that the shielding by the molecular ligands in phthalocyanines might not be sufficient to protect the spin of the metal ion from scattering with conduction electrons of the underlying substrate. Thus, it is reasonable to grow double layers of molecules, where the first layer protects the second one from conduction electrons of the substrate.

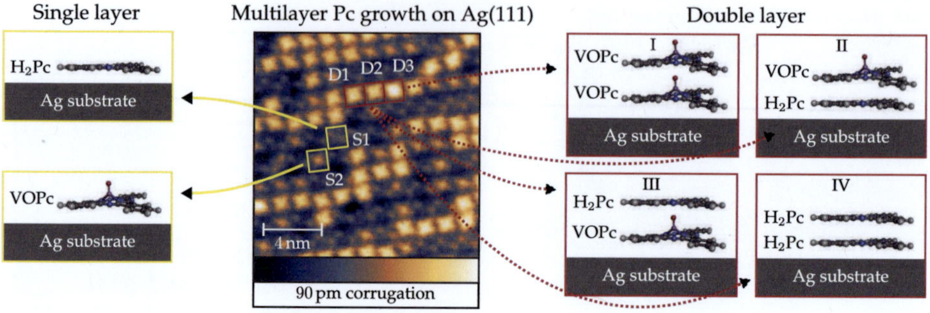

Figure 7.5: *Double layer growth of an H_2Pc-VOPc mixture on Ag(111). The yellow squares identify the two molecule types in the first layer (S1 and S2). At least three different configurations are found in the second layer (D1 - D3). They cannot be clearly assigned to the 4 configurations shown on the right because the underlying layer is not visible. Tunneling parameters:* $U_0 = 600\,\mathrm{mV}$ *and* $I_0 = 300\,\mathrm{pA}$.

7.2.2 Second molecular layer

The growth of an H_2Pc-VOPc mixture in the second molecular layer is much more complicated than in the first layer, as different molecule stacking cannot be distinguished easily. An example is shown in figure 7.5. The STM topography data shows an ordered structure of molecules arranged quadratically. Single (yellow) and double layer coverage (red) is present. The correspondingly colored squares indicate the different situations. In areas where only single layer growth is present, the types of the molecules can, equivalent to the previous section, be identified by the bright dot of the VOPc (S1 and S2). The second layer shows at least three different configurations (D1 - D3). They cannot be clearly attributed to the four different stackings illustrated in the right part of figure 7.5. However, as the configuration D3 is the highest observed, it is assumable that it corresponds to a situation similar to stacking I. The molecules in D1 and D2 are significantly lower than D3 and can therefore be assumed to correspond to situation II or III. It has to be mentioned that this basic characterization does not take lateral displacement nor flipping of the VOPc (VO^{2+}-ion pointing downwards) into account.

Further insight can again be gained by spectroscopic investigations. The observation of a Kondo resonance hints at the presence of a VOPc, but also at insufficient decoupling from the substrate. Thus, all three double layer configurations (D1-D3) have to be compared. This is shown in figure 7.6. Subfigure 7.6***(a)*** shows a magnified view of the double layer situation of figure 7.5. A line-profile (purple) across the molecules confirms the different configurations (fig. 7.6***(b)***).

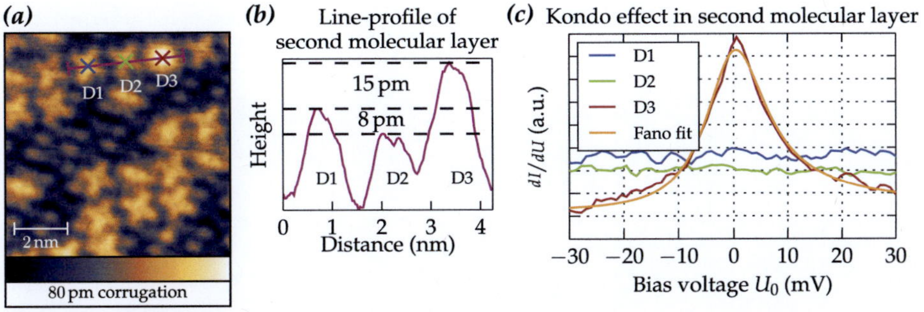

Figure 7.6: Identification of VOPc molecules in the second monolayer. ***(a)****: Enhanced view of the second-monolayer growth of H_2Pc and VOPc on Ag(111). Tunneling parameters: $U_0 = 50\,\mathrm{mV}$ and $I_0 = 320\,\mathrm{pA}$.* ***(b)****: Line-profile as indicated by the purple line in subfigure* ***(a)****.* ***(c)****: Tunneling spectroscopy showing a Fano lineshape, indicating a Kondo effect, only on position D3. The positions D1 and D2 show flat curves. Positions are marked in* ***(a)****.*

In subfigure 7.6*(c)* tunneling spectra of the three different positions marked in part *(a)* are presented. It can be seen that only the highest configuration D3 shows a Fano line-shape at zero energy, indicating a Kondo resonance. The Kondo temperature of $T_K = 79\,\mathrm{K}$ was again obtained by fitting the spectrum of D3 with a Fano function. The decrease of the Kondo temperature in comparison to the first mono-layer is significant as it indicates a lower coupling to the conduction electrons of the substrate. Nevertheless, the presence of the Kondo resonance in D3 indicates insufficient decoupling considering magnetic resonance experiments. This finding supports the assumption of two stacked VOPcs in D3.

However, the two other configurations D1 and D2 both do not show any clear feature. Since D1 is 8 pm higher than D2, it can be concluded that at least one VOPc is present in the stacking of D1. Additionally, the topography slightly suggests that the VOPc is in the upper layer. Unfortunately, the STM tip did not yield a perfect resolution during the study to confirm this conclusion. Nevertheless, a situation is present where preliminary magnetic resonance experiments can be performed as no Kondo resonance was found and most likely a VOPc could be identified in the second layer.

7.2.3 Resonance measurements of VOPc in the second layer

For magnetic resonance measurements a Cr coated W tip was prepared as explained in chapter 4.1.3. The coverage of about 40 ML Cr should yield an

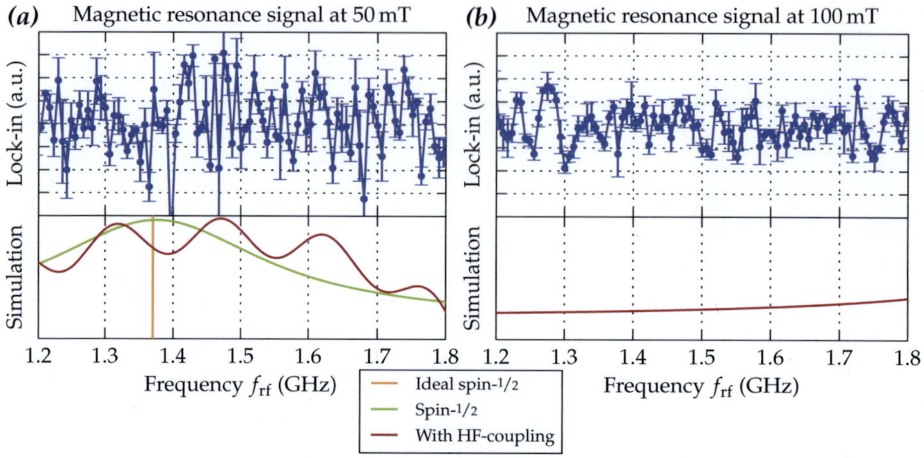

Figure 7.7: *Magnetic resonance measurement and simulations of VOPc molecules.* ***(a)*** *and* ***(b)*** *depict measurements at* 50 *and* 100 mT*, respectively. The blue curves show the average of 5 resonance measurements obtained by stabilizing the tip above a VOPc and sweeping* f_{rf}*. In the lower panels simulations yielding the resonance frequencies of an ideal spin-1/2 (yellow line), a spin-1/2 with broadening (green curve) and a simulation including the hyperfine coupling to V nucleus (red curve) are shown.*

out-of-plane polarization [75]. This is needed as only an out-of-plane magnetic field is available in the setup. The splitting of the states and the polarization of the tip should be parallel to maximize the heterodyne signal according to Green's function calculations of J. Fransson [72]. The use of antiferromagnetic Cr is further necessary to avoid stray fields, which might alter the resonance frequency of the VO^{2+}-ion.

To identify potential resonance peaks, the energy levels obtained by solving the spin Hamiltonian (eq. 7.1), which were shown at the beginning of this chapter (ch. 7.1), can be used to calculate the resonance frequencies of the VO^{2+}-ion. Therefore, equation 2.8 was used and a finite linewidth was assumed.

The simulations together with magnetic resonance measurements are shown in figure 7.7. The spin-polarized STM tip was placed above the stacking D1, which is most likely a VOPc molecule in the second layer. There, the rf-frequency was swept multiple times. These spectra were averaged to increase the signal-to-noise ratio. Subfigure 7.7*(a)* shows a spectrum obtained at 50 mT. A very noisy curve was obtained which shows no clear peaks or trends above the noise limit (blue curve). The simulations at this magnetic field are shown in the lower panel. The frequency of an ideal spin-1/2 is depicted in yellow. The green curve shows a simulation assuming a finite linewidth of a spin-1/2. As the linewidth was was chosen so large that the hyperfine levels cannot be resolved, only a single peak is found. The red curve shows a simulation with a smaller linewidth. Thus, the single resonance peak is split into 8 peaks by the hyperfine coupling, whereof only 4 are visible in the scanned frequency range. It can be seen that none of the simulations matches the spectrum.

Subfigure 7.7*(b)* shows a measurement at the same position, but the magnetic field was increased to 100 mT. Thereby, the resonances should be shifted to a higher frequency and not be observed anymore. This is affirmed by the simulation in the lower panel, where a flat spectrum was obtained as the resonance are much higher at this magnetic field. The blue spectrum in contrast, is barley different from subfigure 7.7*(a)*. Only the noise level is reduced, which might be related to better damping of the STM body at higher external magnetic fields.

In summary, it has to be acknowledged that no evidence for a magnetic resonance signal could be observed in the shown measurements. A significant reason therefore are the advanced tunneling conditions on samples containing molecules: When working with molecules in STM only rather low tunneling currents are possible since high currents imply a very close tip sample distance. Under this conditions, molecules can easily attach to the tip. This was observed regularly through drastic changes of the achieved resolution in STM images. Naturally, such an attached molecule destroys the spin polarization of an STM tip rendering magnetic resonance measurements impossible. Great care was taken to avoid picking up of molecules before the spectra in figure 7.7 were taken. But unfortunately, it cannot be proven that a spin-polarized tip was present during the shown spectra. An approach to improve this circumstance will be shown in the next section.

7.3 *In situ* verification of the spin polarization by Co islands on Cu(111)

7.3.1 The dual deposition technique

A big issue in molecule-based SP-STM experiments is the pollution of the tip with molecules. It was regularly observed that molecules sticked to the tip, suddenly changing the contrast of STM images. This will certainly ruin the ability of spin-polarized studies with this tip. As a spin-polarized tip is crucial for the heterodyne magnetic resonance experiments it must be ensured that no molecules got picked up and magnetic contrast can be obtained with the present tip. For this reason a method for *in situ* verification of the spin polarization had to be found.

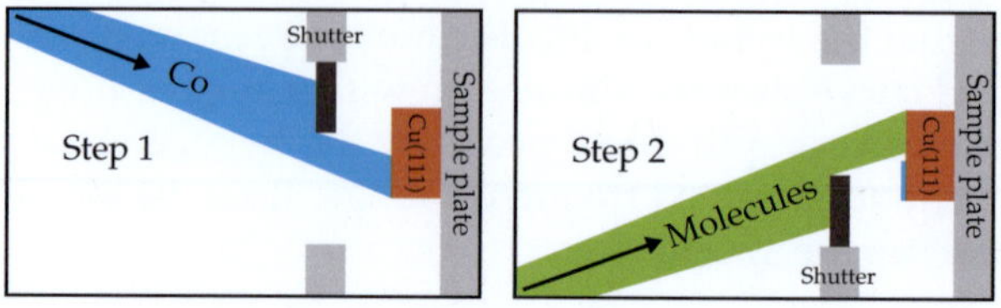

Figure 7.8: Sketch of the subsequent deposition of Co and Pcs onto two separate areas on Cu(111). A small shutter was integrated into the preparation chamber to hide parts of the Cu crystal.

Since the spin polarization of a Cr coated tip with an out-of-plane polarization had to be confirmed, a well-studied system with ferromagnetic domains pointing out-of-plane had to be found. A good sample are Co islands on Cu(111). This system was shown to exhibit double layer islands with out-of-plane spin polarization by Pietzsch *et al.* as early as in 2004 [166]. Thus, the approach is to prepare Co islands and the molecule mixture parallel on one Cu single crystal. Thereby, confirmation of spin polarization and magnetic resonance measurements on the molecules are possible *in situ*.

Unfortunately, molecular double layers and Co island cannot be deposited in the same area of a Cu crystal. The reason for this is that the molecules will preferably diffuse to and upon the Co islands. Hence, the Co islands will be covered with molecules which renders confirmation of the spin polarization of the tip impossible.

To work around this issue the molecules and the Co islands have to be deposited on different sides of the crystal. Thereby, the molecules are macroscopically separated from the Co which prohibits intermixing as the mean free path is not long enough at room temperature. In actual measurements the coarse motion of the STM can be utilized to traverse between the two regions.

The approach is sketched in figure 7.8. As the molecule and the metal evaporator are located at different positions, it was sufficient to install a small shutter in the preparation chamber (ch. 4.1.3). It is located directly in front of the sample and can be rotated by 180°. In the first step Co (blue beam) was deposited in the lower part of the sample. Afterwards, in the second step, the molecular double-layer, identical to the previous section, was prepared (green beam). This procedure yielded excellent samples which allow magnetic resonance studies with *in situ* verification of the spin polarization.

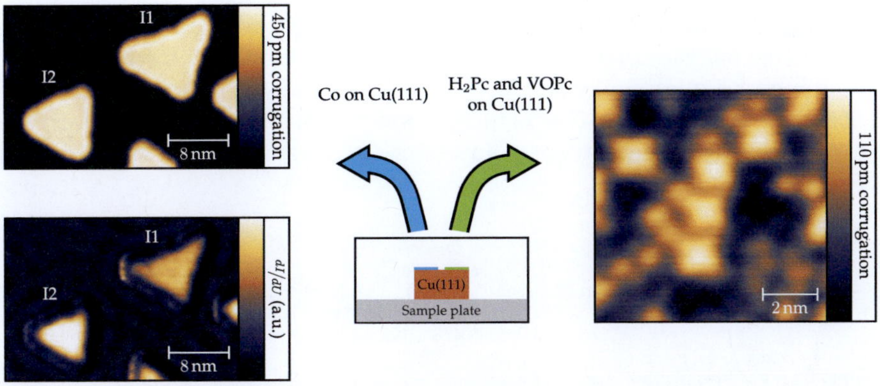

Figure 7.9: *Measurement of a dual sample for confirmation of the spin polarization and magnetic resonance measurements. The sketch in the middle illustrates the two areas on the sample. The left images show triangular Co islands. The spin polarization of the tip could be verified (different* ${}^{dI}/{}_{dU}$ *contrast of island I1 and I2). Tunneling parameters:* $U_0 = -320\,\mathrm{mV}$ *and* $I_0 = 1.5\,\mathrm{nA}$. *On the right side of the sample, VOPc molecules could be identified in the second layer. Tunneling parameters:* $U_0 = 600\,\mathrm{mV}$ *and* $I_0 = 100\,\mathrm{pA}$.

7.3.2 Verification of the spin polarization and magnetic resonance measurements

Following the deposition technique of the previous section, an exemplary sample is shown in figure 7.9. The Cu crystal with its two domains is sketched in the middle. On the left side, triangular Co islands were found. As a Cr coated W tip was used, islands with different magnetization yield different contrasts in local conductance maps (if a bias voltage close to spin polarized state is chosen). This can be clearly seen in the figure as island I2 is brighter than island I1 in the ${}^{dI}/{}_{dU}$-map (lower left image). Thus, the spin polarization of the tip is confirmed. A TMR between 25 and 34 % was achieved.

Subsequently, the tip was retracted, macroscopically moved to the molecule covered area and carefully approached to the surface again. An adequate surface quality, showing VOPc molecules in the second monolayer, was found (right side of figure 7.9). On these molecules, magnetic resonance experiment as in the previous section were performed.

These measurements are shown in figure 7.10, along with corresponding simulations. It can be seen that the spectrum obtained at $50\,\mathrm{mT}$ (blue curve in subfigure ***(a)***) shows a very broad peak. This peak coincides rather well with the simulated green curve. The repeated peaks of the simulation including the hyperfine interaction (red curve), on the other hand, are not reproduced. Immediately after measuring the blue curve in 7.10*(a)*, the magnetic field was increased to 150 mT and the same measurement was repeated. The curve obtained, which does not show the broad peak anymore, is depicted in 7.10*(b)*. This characteristic fits the expected behavior of a magnetic resonance measurement exceedingly well. Unfortunately, a subsequent control measurement at 50 mT did not confirm the peak in subfigure 7.10*(a)*.

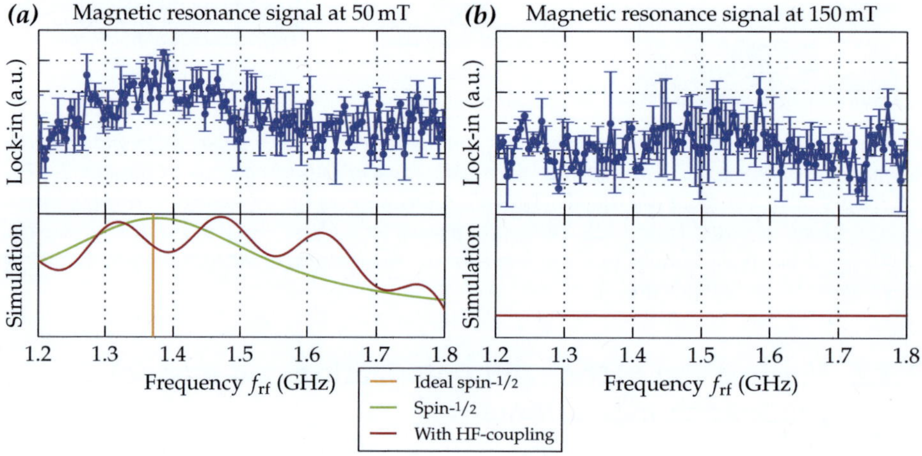

Figure 7.10: *Magnetic resonance measurement of VOPc on Cu(111) after confirmation of the tip spin polarization.* ***(a)*** *and* ***(b)*** *depict measurements at* 50 *and* 150 mT*. The blue curves show the average of 3 resonance measurements obtained by stabilizing the tip above a VOPc and sweeping* f_{rf}*. In the lower panels, simulations yielding the resonance frequencies of an ideal spin-1/2 (yellow line), a spin-1/2 with broadening (green curve) and a simulation including the hyperfine coupling to V nucleus (red curve) are shown.*

This can have various reasons: Firstly, no laterally resolved measurements were performed and no transmission normalization was used, unfortunately. Thus, even slight inaccuracies of the lateral positions of the spectra, which lead to different local density of states, can change the obtained resonance spectra. This argument supports and diminishes the claim of a magnetic resonance signal at once as it cannot be assured that the shown spectra are exactly at the center of the VOPc. Only complete, laterally resolved spectra could clearly proof or falsify the shown resonance-like behavior. Additionally, it must be admitted that, although greatest care was taken, it cannot be guaranteed that the tip state was kept exactly the same before and after the change of the magnetic field.

Nevertheless, a measurement showing very promising spectra of single spin detection in an STM could be presented. This was only possible by optimizing the measurements using *in situ* confirmation of the spin-polarization of the STM tip.

7.4 Summary magnetic resonance of paramagnetic molecules

In the present chapter, preliminary magnetic resonance measurements on paramagnetic molecules were presented. The large variety of paramagnetic molecules requires a wise selection of primary test samples. For this work the phthalocyanines, especially VOPc, were chosen as they are well and widely studied by means of STM.

To ensure a sufficient decoupling from the substrate and to achieve a condition of an "almost" free spin, a rather complicated deposition up to two molecular layers of a molecule mixture was shown. The decoupling could be partially confirmed by spectroscopic investigations concerning the Kondo resonance.

As a tip spin polarization is crucial for the heterodyne magnetic resonance measurements a procedure for *in situ* confirmation of the tip polarization was presented. This allowed magnetic resonance measurements showing signals which agree excellent with spin Hamiltonian simulations. Although, the spectra at different magnetic fields show the awaited behavior, a clear proof is missing as no laterally resolved resonance data could be obtained.

A further experimental modification could be the change to "standard" ESR molecules (DPPH, TEMPO, BDPA). They also exhibit a free radical which should presumably react with the metal surface and vanish in its conduction band. But recent studies claim that the free spin can be retained in some surface-molecule combinations [167–169]. Thus, this could be, along with transmission normalization, a promising area for further studies.

8 Conclusion and outlook

The present PhD thesis focused on the measurement of magnetic resonance in an STM. It was mentioned already in the introduction that this concept can be understood as a combination of two well-established techniques, namely magnetic resonance and STM. The aim was to enable measurements of magnetism dynamics of single nanoscale objects in the GHz range. As this is not possible in conventional STM, a new concept using heterodyne detection was introduced. This concept utilizes spin-polarized tunneling and the application of a radio frequency bias voltage.

Since SP-STM is commonly utilized in the experimental setup, the main challenge throughout this study is the need of a radio frequency modulation of the bias voltage at the tunneling junction. It became clear that this integration yields major difficulties as the electronic situation of an STM is rarely suited for the transmission of these electromagnetic waves. Even after electronic modifications, the primary issue remains the impedance mismatch at the tunneling junction. This mismatch can never be removed and will always cause non-constant transmission as its gigaohm resistance is crucial for STM operation. To tackle this problem a new method for the characterization of the transmission through the setup was introduced. This method is based on the comparison of radiofrequency-based measurements with conventional tunneling spectroscopy measurements. It allows to determine the actual voltage amplitude that has to be produced at the radio frequency generator to achieve a certain voltage amplitude across the tunneling junction. Furthermore, the transmission data obtained can be utilized to normalize the radio frequency voltage and thereby mostly compensating the parasitic modification of the radio frequency voltage. Summarizing, the presented experimental setup was shown to provide the best preconditions possible for initial magnetic resonance measurement after its modification. Only major changes, like the construction of a special coaxial tip-holder [170], could further improve the radio frequency capability of the setup.

The first measurements presented were conducted on vortices in Fe on W(110). As this well-studied system exhibits a gyration mode in the low GHz regime, it is a perfect test sample for the newly proposed technique. Especially the complete micromagnetic simulations allow to precisely predict the expected resonance behavior of a particular island. Thereby, the actual signal of a resonance experiment could be estimated which not only proves the feasibility of the proposed technique but also supports the experimentalist while identifying a resonance signal. Of course, it must be admitted that this first sample cannot fulfill the aim to measure a single spin as the islands are quite large and exhibit a continuous magnetization.

As the initial resonance measurements showed vast signals originating from transmission effects, it became clear that the distinction of the resonance signal with respect to the background is challenging. Two approaches to identify possible resonance signals were presented: Detection of the lateral variation of the resonance behavior and measurements at different external magnetic fields. Both yield hints of possible magnetic resonances inside an Fe island. A big improvement was reached by the utilization of the transmission normalization. Thereby, the acquired spectra could be flattened almost completely. This is advantageous as deviations from the background can be identified more easily. The measurements using this technique yield a very sharp peak which is a promising hint for a resonance signal. But unfortunately, a clear proof by measurements at different magnetic fields is missing. This rather easy next step is of course a promising outlook and can be considered work in progress.

The second sample system introduced were magnetic skyrmions in Fe on Ir(111). This new magnetic system was chosen because an excitation should be possible at lower current density. Unfortunately, only rough approximations of the resonance characteristics are possible. Thus, the broad signature observed in the resonance measurements cannot be clearly attributed to resonances of the skyrmion lattice. A possible future relief could be the simulation of this magnetic structure in the framework of a Heisenberg Hamiltonian. This could allow to confirm the signals observed.

With the prospect of single spin detection, preliminary measurements on paramagnetic molecules were presented. The measurement of the spin precession of an embedded 3d metal ion is an ambiguous idea but could revolutionize the field of surface science. A systematic approach to enhance the decoupling of the spin center from the substrate was presented. This was possible as the excellent lateral resolution of the STM allows to study single molecules even in the second molecular layer. Along with a helpful procedure to confirm the spin polarization of the STM tip, first promising resonance experiments could be presented. They show a broad resonance exactly at the position predicted by basic spin Hamiltonian simulations. Since the influence of an magnetic field on the signal could be confirmed, a strong evidence of a magnetic resonance is present, as well. However, lateral variation of the tip position due to drift cannot be excluded.

In summary, the shown experiments must be considered within the scope of an experimental development. Especially since the new technique to measure magnetic resonance in an STM combines two different research areas. Thus, a solid understanding of the actual signal measured is of greatest importance. It can be seen that the present work carefully developed the expected signals and hence represents a valid basis for this new technique. Although this work could only present hints and not clear proofs of magnetic resonance signals, it is hoped that a stimulating impulse could be given for further experiments regarding magnetic resonance in STM.

Bibliography

[1] G. Binnig *et al*. Surface Studies by Scanning Tunneling Microscopy. *Phys. Rev. Lett.* **49**, 57 (1982).

[2] G. Binnig, H. Fuchs and E. Stoll. Surface diffusion of oxygen atoms individually observed by STM. *Surface Science Letters* **169**, L295–L300 (1986).

[3] M. F. Crommie, C. P. Lutz and D. M. Eigler. Confinement of Electrons to Quantum Corrals on a Metal Surface. *Science* **262**, 218–220 (1993).

[4] J. K. Gimzewski, E. Stoll and R. R. Schlittler. Scanning tunneling microscopy of individual molecules of copper phthalocyanine adsorbed on polycrystalline silver surfaces. *Surface Science* **181**, 267–277 (1987).

[5] P. K. Hansma *et al*. Scanning tunneling microscopy and atomic force microscopy: application to biology and technology. *Science* **242**, 209–216 (1988).

[6] W. Wulfhekel and J. Kirschner. Spin-Polarized Scanning Tunneling Microscopy of Magnetic Structures and Antiferromagnetic Thin Films. *Annual Review of Materials Research* **37**, 69–91 (2007).

[7] R. Wiesendanger. Spin mapping at the nanoscale and atomic scale. *Rev. Mod. Phys.* **81**, 1495–1550 (2009).

[8] C. P. Slichter. *Principles of magnetic resonance* (1996).

[9] R. P. Feynman. Simulating physics with computers. *Int J Theor Phys* **21**, 467–488 (1982).

[10] D. Loss and D. P. DiVincenzo. Quantum computation with quantum dots. *Phys. Rev. A* **57**, 120–126 (1998).

[11] L. Bogani and W. Wernsdorfer. Molecular spintronics using single-molecule magnets. *Nat Mater* **7**, 179–186 (2008).

[12] T. Miyamachi *et al*. Stabilizing the magnetic moment of single holmium atoms by symmetry. *Nature* **503**, 242–246 (2013).

[13] I. L. Chuang *et al*. Experimental realization of a quantum algorithm. *Nature* **393**, 143–146 (1998).

[14] L. M. K. Vandersypen *et al.* Experimental realization of Shor's quantum factoring algorithm using nuclear magnetic resonance. *Nature* **414**, 883–887 (2001).

[15] E. Zavoisky. Paramagnetic Absorption in Some Salts in Perpendicular Magnetic Fields. *Zhurnal Eksperimentalnoi Teor. Fiz.* **16**, 603–606 (1946).

[16] J. Griffiths. Anomalous High-Frequency Reistance of Ferromagnetic Metals. *Nature* **158**, 670–671 (1946).

[17] M. Drescher, N. Kaplan and E. Dormann. Current density imaging by pulsed conduction electron spin resonance. *Journal of Magnetic Resonance* **184**, 44–50 (2007).

[18] M. N. Leuenberger and D. Loss. Quantum computing in molecular magnets. *Nature* **410**, 789–793 (2001).

[19] F. Bloch. Nuclear Induction. *Phys. Rev.* **70**, 460–474 (1946).

[20] H. Haken and H. C. Wolf. *Atom- und Quantenphysik : Einführung in die experimentellen und theoretischen Grundlagen* (2004).

[21] J. J. Sakurai and S. F. H. Tuan. *Modern quantum mechanics* (2008).

[22] W. Heisenberg. Über den anschaulichen Inhalt der quantentheoretischen Kinematik und Mechanik. *Z. Physik* **43**, 172–198 (1927).

[23] E. L. Hahn. Spin Echoes. *Phys. Rev.* **80**, 580–594 (1950).

[24] F. Schwabl. *Quantenmechanik (QM I) : Eine Einführung* (2007).

[25] H. Kramers. Théorie générale de la rotation paramagnétique dans les cristaux. *Proceedings of the Koninklijke Akademie Van Wetenschappen Te Amsterdam* **33**, 959 (1930).

[26] M. J. Klein. On a Degeneracy Theorem of Kramers. *American Journal of Physics* **20**, 65 (1952).

[27] B. Bleaney and K. W. H. Stevens. Paramagnetic resonance. *Rep. Prog. Phys.* **16**, 108 (1953).

[28] W. Heisenberg. Mehrkörperproblem und Resonanz in der Quantenmechanik. *Z. Phys.* **38**, 411 (1926).

[29] I. Dzyaloshinsky. A thermodynamic theory of "weak" ferromagnetism of antiferromagnetics. *Journal of Physics and Chemistry of Solids* **4**, 241–255 (1958).

[30] T. Moriya. Anisotropic Superexchange Interaction and Weak Ferromagnetism. *Phys. Rev.* **120**, 91–98 (1960).

[31] A. N. Bogdanov and U. K. Rößler. Chiral Symmetry Breaking in Magnetic Thin Films and Multilayers. *Phys. Rev. Lett.* **87**, 037203 (2001).

[32] S. V. H. Vonsovskij. *Ferromagnetic resonance : the phenomenon of resonance absorption of hf electromagnetic field energy in ferromagnetic materials* (1964).

[33] L. D. Landau and E. M. Lifshitz. On the theory of the dispersion of magnetic permeability in ferromagnetic bodies. *Phys. Zeitsch. der Sow.* **8**, 153–169 (1935).

[34] T. Gilbert. A phenomenological theory of damping in ferromagnetic materials. *IEEE Transactions on Magnetics* **40**, 3443–3449 (2004).

[35] R. K. Wangsness. Magnetic Resonance and Minimum Entropy Production. Macroscopic Equations of Motion. *Phys. Rev.* **104**, 857–866 (1956).

[36] H. Suhl. Ferromagnetic Resonance in Nickel Ferrite Between One and Two Kilomegacycles. *Phys. Rev.* **97**, 555–557 (1955).

[37] C. Kittel. On the Theory of Ferromagnetic Resonance Absorption. *Phys. Rev.* **73**, 155–161 (1948).

[38] M. T. Weiss and P. W. Anderson. Ferromagnetic Resonance in Ferroxdure. *Phys. Rev.* **98**, 925–926 (1955).

[39] E. Hecht. *Optik* (2009).

[40] T. A. Klar and S. W. Hell. Subdiffraction resolution in far-field fluorescence microscopy. *Opt. Lett.* **24**, 954–956 (1999).

[41] L. d. Broglie. The reinterpretation of wave mechanics. *Found Phys* **1**, 5–15 (1970).

[42] J. Bardeen. Tunnelling from a Many-Particle Point of View. *Phys. Rev. Lett.* **6**, 57–59 (1961).

[43] C. J. Chen. *Introduction to scanning tunneling microscopy* (2008).

[44] J. Tersoff and D. R. Hamann. Theory and Application for the Scanning Tunneling Microscope. *Phys. Rev. Lett.* **50**, 1998–2001 (1983).

[45] J. Tersoff and D. R. Hamann. Theory of the scanning tunneling microscope. *Phys. Rev. B* **31**, 805–813 (1985).

[46] R. Meservey and P. M. Tedrow. Spin-polarized electron tunneling. *Physics Reports* **238**, 173–243 (1994).

[47] M. Eltschka *et al.* Probing Absolute Spin Polarization at the Nanoscale. *Nano Lett.* **14**, 7171–7174 (2014).

[48] D. T. Pierce. Spin-polarized electron microscopy. *Phys. Scr.* **38**, 291 (1988).

[49] S. F. Alvarado and P. Renaud. Observation of spin-polarized-electron tunneling from a ferromagnet into GaAs. *Phys. Rev. Lett.* **68**, 1387–1390 (1992).

[50] M. Julliere. Tunneling between ferromagnetic films. *Physics Letters A* **54**, 225–226 (1975).

[51] T. Miyazaki and N. Tezuka. Giant magnetic tunneling effect in Fe/Al2O3/Fe junction. *Journal of Magnetism and Magnetic Materials* **139**, L231–L234 (1995).

[52] R. Wiesendanger *et al.* Observation of vacuum tunneling of spin-polarized electrons with the scanning tunneling microscope. *Phys. Rev. Lett.* **65**, 247–250 (1990).

[53] M. Bode, M. Getzlaff and R. Wiesendanger. Spin-Polarized Vacuum Tunneling into the Exchange-Split Surface State of Gd(0001). *Phys. Rev. Lett.* **81**, 4256–4259 (1998).

[54] A. Wachowiak *et al.* Direct Observation of Internal Spin Structure of Magnetic Vortex Cores. *Science* **298**, 577–580 (2002).

[55] A. Yamasaki *et al.* Direct Observation of the Single-Domain Limit of Fe Nanomagnets by Spin-Polarized Scanning Tunneling Spectroscopy. *Phys. Rev. Lett.* **91**, 127201 (2003).

[56] M. A. Hemminga and L. J. Berliner. *ESR Spectroscopy in Membrane Biophysics* (2007).

[57] O. Y. Grinberg and L. J. Berliner. *Very High Frequency (VHF) ESR/EPR* (2004).

[58] T. Balashov *et al.* Magnetic Anisotropy and Magnetization Dynamics of Individual Atoms and Clusters of Fe and Co on Pt(111). *Phys. Rev. Lett.* **102**, 257203 (2009).

[59] N. Tsukahara *et al.* Adsorption-Induced Switching of Magnetic Anisotropy in a Single Iron(II) Phthalocyanine Molecule on an Oxidized Cu(110) Surface. *Phys. Rev. Lett.* **102**, 167203 (2009).

[60] A. A. Khajetoorians *et al.* Itinerant Nature of Atom-Magnetization Excitation by Tunneling Electrons. *Phys. Rev. Lett.* **106**, 037205 (2011).

[61] S. Loth *et al.* Bistability in Atomic-Scale Antiferromagnets. *Science* **335**, 196–199 (2012).

[62] A. V. Balatsky, Y. Manassen and R. Salem. ESR-STM of a single precessing spin: Detection of exchange-based spin noise. *Phys. Rev. B* **66**, 195416 (2002).

[63] A. V. Balatsky, M. Nishijima and Y. Manassen. Electron spin resonance-scanning tunneling microscopy. *Advances in Physics* **61**, 117–152 (2012).

[64] Y. Manassen *et al.* Direct observation of the precession of individual paramagnetic spins on oxidized silicon surfaces. *Phys. Rev. Lett.* **62**, 2531 (1989).

[65] Y. Manassen *et al.* Electron spin resonance-scanning tunneling microscopy experiments on thermally oxidized Si(111). *Phys. Rev. B* **48**, 4887 (1993).

[66] P. Messina *et al.* Spin noise fluctuations from paramagnetic molecular adsorbates on surfaces. *Journal of Applied Physics* **101**, 053916–053916–14 (2007).

[67] J. Fransson, O. Eriksson and A. V. Balatsky. Theory of spin-polarized scanning tunneling microscopy applied to local spins. *Phys. Rev. B* **81**, 115454 (2010).

[68] C. Durkan and M. E. Welland. Electronic spin detection in molecules using scanning-tunneling- microscopy-assisted electron-spin resonance. *Appl. Phys. Lett.* **80**, 458 (2002).

[69] G. Nunes and M. R. Freeman. Picosecond Resolution in Scanning Tunneling Microscopy. *Science* **262**, 1029–1032 (1993).

[70] S. Weiss *et al.* Ultrafast scanning probe microscopy. *Applied Physics Letters* **63**, 2567–2569 (1993).

[71] S. Loth *et al.* Measurement of Fast Electron Spin Relaxation Times with Atomic Resolution. *Science* **329**, 1628 –1630 (2010).

[72] J. Fransson and W. Wulfhekel. private communication (2010).

[73] T. Kokubo, A. Gallagher and J. L. Hall. Optical heterodyne detection at a silver scanning tunneling microscope junction. *Journal of Applied Physics* **85**, 1311–1316 (1999).

[74] E. Matsuyama *et al.* Principles and Application of Heterodyne Scanning Tunnelling Spectroscopy. *Sci. Rep.* **4** (2014).

[75] M. Bode. Spin-polarized scanning tunnelling microscopy. *Rep. Prog. Phys.* **66**, 523 (2003).

[76] J. C. Slonczewski. Current-driven excitation of magnetic multilayers. *Journal of Magnetism and Magnetic Materials* **159**, L1–L7 (1996).

[77] L. Berger. Emission of spin waves by a magnetic multilayer traversed by a current. *Phys. Rev. B* **54**, 9353–9358 (1996).

[78] E. B. Myers *et al.* Current-Induced Switching of Domains in Magnetic Multilayer Devices. *Science* **285**, 867–870 (1999).

[79] J. Kim *et al.* A 45nm 1Mb embedded STT-MRAM with design techniques to minimize read-disturbance. In *2011 Symposium on VLSI Circuits (VLSIC)*, pages 296–297 (2011).

[80] S. Zhang, P. M. Levy and A. Fert. Mechanisms of Spin-Polarized Current-Driven Magnetization Switching. *Phys. Rev. Lett.* **88**, 236601 (2002).

[81] S. Zhang and Z. Li. Roles of Nonequilibrium Conduction Electrons on the Magnetization Dynamics of Ferromagnets. *Phys. Rev. Lett.* **93**, 127204 (2004).

[82] A. Kubetzka *et al.* Spin-Polarized Scanning Tunneling Microscopy with Antiferromagnetic Probe Tips. *Phys. Rev. Lett.* **88**, 057201 (2002).

[83] A. A. Tulapurkar *et al.* Spin-torque diode effect in magnetic tunnel junctions. *Nature* **438**, 339–342 (2005).

[84] T. Balashov. *Inelastic scanning tunneling spectroscopy: magnetic excitations on the nanoscale*. Ph.D. thesis (2009).

[85] L. Gerhard *et al.* Magnetoelectric coupling at metal surfaces. *Nat Nano* **5**, 792–797 (2010).

[86] S. Schmaus *et al.* Giant magnetoresistance through a single molecule. *Nat Nano* **6**, 185–189 (2011).

[87] S. Holzberger *et al.* Parity Effect in the Ground state Localization of Antiferromagnetic Chains Coupled to a Ferromagnet. *Phys. Rev. Lett.* **110**, 157206 (2013).

[88] A. Sonntag *et al.* Thermal Stability of an Interface-Stabilized Skyrmion Lattice. *Phys. Rev. Lett.* **113**, 077202 (2014).

[89] R. Musket *et al.* Preparation of atomically clean surfaces of selected elements: A review. *Applications of Surface Science* **10**, 143–207 (1982).

[90] D. M. Pozar. *Microwave engineering* (2012).

[91] A. Hubert and R. Schäfer. *Magnetic domains : the analysis of magnetic microstructures* (1998).

[92] G. Abo *et al.* Definition of Magnetic Exchange Length. *IEEE Transactions on Magnetics* **49**, 4937–4939 (2013).

[93] K. Y. Guslienko. Magnetic Vortex State Stability, Reversal and Dynamics in Restricted Geometries. *Journal of Nanoscience and Nanotechnology* **8**, 2745–2760 (2008).

[94] B. E. Argyle, E. Terrenzio and J. C. Slonczewski. Magnetic Vortex Dynamics Using the Optical Cotton-Mouton Effect. *Phys. Rev. Lett.* **53**, 190–193 (1984).

[95] R. P. Cowburn *et al.* Single-Domain Circular Nanomagnets. *Phys. Rev. Lett.* **83**, 1042–1045 (1999).

[96] T. Shinjo *et al.* Magnetic Vortex Core Observation in Circular Dots of Permalloy. *Science* **289**, 930–932 (2000).

[97] A. A. Thiele. Steady-State Motion of Magnetic Domains. *Phys. Rev. Lett.* **30**, 230–233 (1973).

[98] G. M. Wysin. Magnetic vortex mass in two-dimensional easy-plane magnets. *Phys. Rev. B* **54**, 15156–15162 (1996).

[99] B. A. Ivanov *et al.* Magnon modes and magnon-vortex scattering in two-dimensional easy-plane ferromagnets. *Phys. Rev. B* **58**, 8464–8474 (1998).

[100] K. Y. Guslienko *et al.* Eigenfrequencies of vortex state excitations in magnetic submicron-size disks. *Journal of Applied Physics* **91**, 8037–8039 (2002).

[101] D. L. Huber. Dynamics of spin vortices in two-dimensional planar magnets. *Phys. Rev. B* **26**, 3758–3765 (1982).

[102] V. Novosad *et al.* Spin excitations of magnetic vortices in ferromagnetic nanodots. *Phys. Rev. B* **66**, 052407 (2002).

[103] S.-B. Choe *et al.* Vortex Core-Driven Magnetization Dynamics. *Science* **304**, 420–422 (2004).

[104] M. Bode *et al.* Thickness dependent magnetization states of Fe islands on W(110): From single domain to vortex and diamond patterns. *Applied Physics Letters* **84**, 948–950 (2004).

[105] L. Berbil-Bautista. *Ferromagnetic thin-films and nanostructures studied by spin-polarized scanning tunneling microscopy*. Ph.D. thesis (2006).

[106] N. Rougemaille and A. K. Schmid. Self-organization and magnetic domain microstructure of Fe nanowire arrays. *Journal of Applied Physics* **99**, 08S502 (2006).

[107] H.-J. Elmers, J. Hauschild and U. Gradmann. Critical behavior of the uniaxial ferromagnetic monolayer Fe(110) on W(110). *Phys. Rev. B* **54**, 15224–15233 (1996).

[108] H. J. Elmers, J. Hauschild and U. Gradmann. Onset of perpendicular magnetization in nanostripe arrays of Fe on stepped W(110) surfaces. *Phys. Rev. B* **59**, 3688–3695 (1999).

[109] A. Wachowiak. *Aufbau einer 300mK-Ultrahochvakuum-Rastertunnelmikroskopie-Anlage mit 14 Tesla Magnet und spinpolarisierte Rastertunnelspektroskopie an ferromagnetischen Fe-Inseln*. Ph.D. thesis (2003).

[110] B. Yang. *Numerical studies of dynamical micromagnetics*. Ph.D. thesis (1997).

[111] T. Fischbacher *et al.* A Systematic Approach to Multiphysics Extensions of Finite-Element-Based Micromagnetic Simulations: Nmag. *IEEE Transactions on Magnetics* **43**, 2896–2898 (2007).

[112] J. Schöberl. NETGEN An advancing front 2D/3D-mesh generator based on abstract rules. *Comput Visual Sci* **1**, 41–52 (1997).

[113] M. J. Donahue and R. D. McMichael. Exchange energy representations in computational micromagnetics. *Physica B: Condensed Matter* **233**, 272–278 (1997).

[114] M. J. Donahue. A variational approach to exchange energy calculations in micromagnetics. *Journal of Applied Physics* **83**, 6491–6493 (1998).

[115] M. J. Donahue and D. G. Porter. Exchange energy formulations for 3D micromagnetics. *Physica B: Condensed Matter* **343**, 177–183 (2004).

[116] U. Gradmann. Magnetic surface anisotropies. *Journal of Magnetism and Magnetic Materials* **54–57, Part 2**, 733–736 (1986).

[117] H. J. Elmers and U. Gradmann. Magnetic anisotropies in Fe(110) films on W(110). *Appl. Phys. A* **51**, 255–263 (1990).

[118] U. Gradmann *et al.* The ferromagnetic monolayer Fe(110) on W(110). *Hyperfine Interact* **57**, 1845–1858 (1990).

[119] W. Nolting. *Klassische Mechanik* (2013).

[120] S. Heinze *et al.* Spontaneous atomic-scale magnetic skyrmion lattice in two dimensions. *Nat Phys* **7**, 713–718 (2011).

[121] T. H. R. Skyrme. A Non-Linear Theory of Strong Interactions. *Proceedings of the Royal Society of London A: Mathematical, Physical and Engineering Sciences* **247**, 260–278 (1958).

[122] T. H. R. Skyrme. A Unified Model of K- and π-mesons. *Proceedings of the Royal Society of London A: Mathematical, Physical and Engineering Sciences* **252**, 236–245 (1959).

[123] T. H. R. Skyrme. A Non-Linear Field Theory. *Proceedings of the Royal Society of London A: Mathematical, Physical and Engineering Sciences* **260**, 127–138 (1961).

[124] T. H. R. Skyrme. A unified field theory of mesons and baryons. *Nuclear Physics* **31**, 556–569 (1962).

[125] A. N. Bogdanov and D. Yablonskii. Thermodynamically stable "vortices" in magnetically ordered crystals. The mixed state of magnets. *Journal of Experimental and Theoretical Physics* **95**, 178 (1989).

[126] A. Bogdanov. New localized solutions of the nonlinear field equations. *Soviet Journal of Experimental and Theoretical Physics Letters* **62**, 247 (1995).

[127] K. Everschor. *Current-Induced Dynamics of Chiral Magnetic Structures : Skyrmions, Emergent Electrodynamics and Spin-Transfer Torques*. Ph.D. thesis (2012).

[128] A. Fert, V. Cros and J. Sampaio. Skyrmions on the track. *Nat Nano* **8**, 152–156 (2013).

[129] A. Fert and P. M. Levy. Role of Anisotropic Exchange Interactions in Determining the Properties of Spin-Glasses. *Phys. Rev. Lett.* **44**, 1538–1541 (1980).

[130] X. Z. Yu *et al.* Real-space observation of a two-dimensional skyrmion crystal. *Nature* **465**, 901–904 (2010).

[131] S. Mühlbauer *et al.* Skyrmion Lattice in a Chiral Magnet. *Science* **323**, 915–919 (2009).

[132] U. K. Rößler, A. N. Bogdanov and C. Pfleiderer. Spontaneous skyrmion ground states in magnetic metals. *Nature* **442**, 797–801 (2006).

[133] S. Andrieu, M. Piecuch and J. F. Bobo. Fe growth on (0001) hcp Ru and (111) fcc Ir: Consequences for structural and magnetic properties. *Phys. Rev. B* **46**, 4909–4916 (1992).

[134] W. L. O'Brien and B. P. Tonner. Magnetic and structural properties of ultrathin Mn and Fe films on Ir(111). *Journal of Vacuum Science & Technology A* **13**, 1544–1548 (1995).

[135] K. von Bergmann *et al.* Observation of a Complex Nanoscale Magnetic Structure in a Hexagonal Fe Monolayer. *Phys. Rev. Lett.* **96**, 167203 (2006).

[136] K. v. Bergmann *et al.* Complex magnetism of the Fe monolayer on Ir(111). *New J. Phys.* **9**, 396 (2007).

[137] M. Menzel. *Non-collinear magnetic ground states observed in iron nanostructures on iridium surfaces*. Ph.D. thesis (2011).

[138] K. von Bergmann *et al.* Tunneling anisotropic magnetoresistance on the atomic scale. *Phys. Rev. B* **86**, 134422 (2012).

[139] C. Gould *et al.* Tunneling Anisotropic Magnetoresistance: A Spin-Valve-Like Tunnel Magnetoresistance Using a Single Magnetic Layer. *Phys. Rev. Lett.* **93**, 117203 (2004).

[140] F. Jonietz *et al.* Spin Transfer Torques in MnSi at Ultralow Current Densities. *Science* **330**, 1648–1651 (2010).

[141] K. Everschor *et al.* Current-induced rotational torques in the skyrmion lattice phase of chiral magnets. *Phys. Rev. B* **84**, 064401 (2011).

[142] K. Everschor *et al.* Rotating skyrmion lattices by spin torques and field or temperature gradients. *Phys. Rev. B* **86**, 054432 (2012).

[143] J. Iwasaki, M. Mochizuki and N. Nagaosa. Universal current-velocity relation of skyrmion motion in chiral magnets. *Nat Commun* **4**, 1463 (2013).

[144] J. Sampaio *et al.* Nucleation, stability and current-induced motion of isolated magnetic skyrmions in nanostructures. *Nat Nano* **8**, 839–844 (2013).

[145] M. Mochizuki. Spin-Wave Modes and Their Intense Excitation Effects in Skyrmion Crystals. *Phys. Rev. Lett.* **108**, 017601 (2012).

[146] Y. Onose *et al.* Observation of Magnetic Excitations of Skyrmion Crystal in a Helimagnetic Insulator Cu_2OSeO_3. *Phys. Rev. Lett.* **109**, 037603 (2012).

[147] S. Seki *et al.* Observation of Skyrmions in a Multiferroic Material. *Science* **336**, 198–201 (2012).

[148] J. T. Grant. Some studies on the Ir(111) surface using LEED and Auger electron spectroscopy. *Surface Science* **25**, 451–456 (1971).

[149] H. P. Singh. Determination of thermal expansion of germanium, rhodium and iridium by X-rays. *Acta Crystallographica Section A* **24**, 469–471 (1968).

[150] N. Knorr *et al.* Kondo Effect of Single Co Adatoms on Cu Surfaces. *Phys. Rev. Lett.* **88**, 096804 (2002).

[151] U. G. E. Perera *et al.* Spatially Extended Kondo State in Magnetic Molecules Induced by Interfacial Charge Transfer. *Phys. Rev. Lett.* **105**, 106601 (2010).

[152] T. Komeda *et al.* Observation and electric current control of a local spin in a single-molecule magnet. *Nat Commun* **2**, 217 (2011).

[153] G. Löbbert. Phthalocyanines. In *Ullmann's Encyclopedia of Industrial Chemistry*. Wiley-VCH Verlag GmbH & Co. KGaA (2000).

[154] C. E. Dent, R. P. Linstead and A. R. Lowe. 217. Phthalocyanines. Part VI. The structure of the phthalocyanines. *J. Chem. Soc.* pages 1033–1039 (1934).

[155] J. M. Robertson. 255. An X-ray study of the phthalocyanines. Part II. Quantitative structure determination of the metal-free compound. *J. Chem. Soc.* pages 1195–1209 (1936).

[156] S. Schmaus. *Spintronics with individual metal-organic molecules.* Ph.D. thesis (2011).

[157] A. B. P. Lever. 336. The magnetic behaviour of transition-metal phathalocyanines. *J. Chem. Soc.* pages 1821–1829 (1965).

[158] J. M. Assour, J. Goldmacher and S. E. Harrison. Electron Spin Resonance of Vanadyl Phthalocyanine. *The Journal of Chemical Physics* **43**, 159–165 (1965).

[159] T. G. Gopakumar *et al.* Coverage-Driven Electronic Decoupling of Fe-Phthalocyanine from a Ag(111) Substrate. *J. Phys. Chem. C* **115**, 12173–12179 (2011).

[160] G. M. Delacote, J. P. Fillard and F. J. Marco. Electron injection in thin films of copper phtalocyanine. *Solid State Communications* **2**, 373–376 (1964).

[161] T. Basova *et al.* Volatile Phthalocyanines: Vapor Pressure and Thermodynamics. *Critical Reviews in Solid State and Materials Sciences* **34**, 180–189 (2009).

[162] P. W. Anderson. Localized Magnetic States in Metals. *Phys. Rev.* **124**, 41–53 (1961).

[163] U. Fano. Effects of Configuration Interaction on Intensities and Phase Shifts. *Phys. Rev.* **124**, 1866–1878 (1961).

[164] V. Madhavan *et al.* Tunneling into a Single Magnetic Atom: Spectroscopic Evidence of the Kondo Resonance. *Science* **280**, 567–569 (1998).

[165] O. Újsághy *et al.* Theory of the Fano Resonance in the STM Tunneling Density of States due to a Single Kondo Impurity. *Phys. Rev. Lett.* **85**, 2557–2560 (2000).

[166] O. Pietzsch *et al.* Spin-Polarized Scanning Tunneling Spectroscopy of Nanoscale Cobalt Islands on Cu(111). *Phys. Rev. Lett.* **92**, 057202 (2004).

[167] P. Krukowski *et al.* Formation of dense nitroxide radical layers on the Au(1 1 1) substrate for ESN-STM measurement. *Applied Surface Science* **255**, 1921–1928 (2008).

[168] P. Krukowski *et al.* Self-assembled monolayers of radical molecules physisorbed on HOPG(0 0 0 1) substrate studied by scanning tunnelling microscopy and electron paramagnetic resonance techniques. *Applied Surface Science* **255**, 8769–8773 (2009).

[169] M. Mas-Torrent *et al.* Attaching Persistent Organic Free Radicals to Surfaces: How and Why. *Chem. Rev.* **112**, 2506–2527 (2012).

[170] C. Saunus *et al.* Versatile scanning tunneling microscopy with 120 ps time resolution. *Applied Physics Letters* **102**, 051601 (2013).

Acknowledgments

- I thank **Prof. Dr. W. Wulfhekel** for being my doctorate supervisor and for giving me the chance to work on this challenging project. I thank him for his trust in my abilities, for his boundless optimism and the motivation he spreads. His unique personal mentoring helped and taught me at once.
- I also want to thank **Prof. Dr. B. Pilawa** for agreeing to be the second referee and for investing time and effort to read this work.
- I specially thank **Dr. Marie Hervé**. Without her help this work would not have reached this point. I thank her for the cooperative work in the lab, for proof reading this thesis and for her merry mind. I think we made a very good team!
- I also want to thank **Lorenz Schmidt** for the support during the work on magnetic resonance in molecules.
- I thank the group of Prof. Dr. A. Ustinov, especially **Sebastian Probst**, for the help with many rf-circuit problems and the loan of some electronics for test purposes.
- I thank **Prof. Dr. Chien-Cheng Kuo** for giving me the opportunity to join his research in Taiwan. And of course I thank **Shih-Yu Wu** for fantastic times, also beside physics.
- I want to thank **Martin Bowen**, **Eric Beaurepaire** and **Manuel Gruber** for the fruitful cooperation and the nice time near Paris.
- I thank all other, present and former, members of the **Wulfhekel group**. I enjoyed the friendly atmosphere during work and afterwards. Almost all students and of course the long term colleagues **Michael, Lukas, Tobias M., Jasmin, Jinjie, Tobias S., Lei, Toshio, Timofey** and **Hironari** contributed to this atmosphere which supported me to undertake this venture.
- I thank the **KHYS** (Karlsruhe House of Young Scientists) for financial support through the Landesgraduiertenförderung (LGF).
- I thank **Lena Hoffmann** for proof-reading this thesis and antagonizing many of my spelling mistakes.
- I want to thank my **parents** and my **siblings** for their support.
- And I thank **Anni** for her unshakable belief in me and even more important for her love 🐵❤️🐴.

Experimental Condensed Matter Physics
(ISSN 2191-9925)

Herausgeber
Physikalisches Institut

Prof. Dr. Hilbert von Löhneysen
Prof. Dr. Alexey Ustinov
Prof. Dr. Georg Weiß
Prof. Dr. Wulf Wulfhekel

Die Bände sind unter www.ksp.kit.edu als PDF frei verfügbar
oder als Druckausgabe bestellbar.

Band 1 Alexey Feofanov
Experiments on flux qubits with pi-shifters. 2011
ISBN 978-3-86644-644-1

Band 2 Stefan Schmaus
Spintronics with individual metal-organic molecules. 2011
ISBN 978-3-86644-649-6

Band 3 Marc Müller
Elektrischer Leitwert von magnetostriktiven Dy-Nanokontakten. 2011
ISBN 978-3-86644-726-4

Band 4 Torben Peichl
Einfluss mechanischer Deformation auf atomare Tunnelsysteme – untersucht mit Josephson Phasen-Qubits. 2012
ISBN 978-3-86644-837-7

Band 5 Dominik Stöffler
Herstellung dünner metallischer Brücken durch Elektromigration und Charakterisierung mit Rastersondentechniken. 2012
ISBN 978-3-86644-843-8

Band 6 Tihomir Tomanic
Untersuchung des elektronischen Oberflächenzustands von Ag-Inseln auf supraleitendem Niob (110). 2012
ISBN 978-3-86644-898-8

Band 7 Lukas Gerhard
Magnetoelectric coupling at metal surfaces. 2013
ISBN 978-3-7315-0063-6

Experimental Condensed Matter Physics
(ISSN 2191-9925)

Band 8 Kirill Fedorov
Fluxon readout for superconducting flux qubits. 2013
ISBN 978-3-7315-0067-4

Band 9 Jochen Zimmer
Cooper pair transport in arrays of Josephson junctions. 2014
ISBN 978-3-7315-0130-5

Band 10 Oliver Berg
Elektrischer Transport durch Nanokontakte von Selten-Erd-Metallen. 2014
ISBN 978-3-7315-0209-8

Band 11 Grigorij Jur'evic Grabovskij
Investigation of coherent microscopic defects inside the tunneling barrier of a Josephson junction. 2014
ISBN 978-3-7315-0210-4

Band 12 Cornelius Thiele
STM Characterization of Phenylene-Ethynylene Oligomers on Au(111) and their Integration into Carbon Nanotube Nanogaps. 2014
ISBN 978-3-7315-0235-7

Band 13 Michael Peter Schackert
Scanning Tunneling Spectroscopy on Electron-Boson Interactions in Superconductors. 2014
ISBN 978-3-7315-0238-8

Band 14 Susanne Butz
One-Dimensional Tunable Josephson Metamaterials. 2014
ISBN 978-3-7315-0271-5

Band 15 Philipp Jung
Nonlinear Effects in Superconducting Quantum Interference Meta-Atoms. 2014
ISBN 978-3-7315-0294-4

Band 16 Sebastian Probst
Hybrid quantum system based on rare earth doped crystals. 2015
ISBN 978-3-7315-0345-3

Experimental Condensed Matter Physics
(ISSN 2191-9925)

Band 17 Wolfram Kittler
Magnetische Anisotropie und Quantenphasenübergang in $CeTi_{1-x}V_xGe_3$. 2015
ISBN 978-3-7315-0363-7

Band 18 Moritz Peter
Towards magnetic resonance in scanning tunneling microscopy using heterodyne detection. 2015
ISBN 978-3-7315-0410-8